# Rolf Sattler

# Biophilosophy

## Analytic and Holistic Perspectives

With 27 Figures

Springer-Verlag
Berlin Heidelberg New York Tokyo

Dr. ROLF SATTLER
Biology Department
McGill University
1205 Ave. Docteur Penfield
Montreal, Quebec H3A 1B1
Canada

ISBN 3-540-16418-9 Springer-Verlag Berlin Heidelberg New York Tokyo
ISBN 0-387-16418-9 Springer-Verlag New York Heidelberg Berlin Tokyo

Library of Congress Cataloging-in-Publication Data. Sattler, R. (Rolf) Biophilosophy: analytic and holistic perspectives. Includes index. 1. Biology – Philosophy. I. Title. QH331.S255   1986   574'.01   86-6699

© by Springer-Verlag Berlin Heidelberg 1986
Printed in Germany

Typesetting, offsetprinting and bookbinding: Brühlsche Universitätsdruckerei, Giessen
2131/3130-543210

*To Hyacinth,*
*beautiful woman and flower*

# Preface

This book is an introduction to biophilosophy, written primarily
for the student of biology, the practicing biologist, and the educated
layperson. It does not presuppose technical knowledge in biology
or philosophy. However, it requires a willingness to examine the
most basic foundations of biology which are so often taken for
granted. Furthermore, it points to the bottomlessness of these
foundations, the mystery of life, the Unnamable ...

I have tried to further the awareness that biological statements
are based on philosophical assumptions which are present in our
minds even before we enter the laboratory. These assumptions,
which often harbor strong commitments, are exposed throughout
the book. I have tried to show how they influence concrete biological research as well as our personal existence and society. Thus,
emphasis is placed on the connection between biophilosophy and
biological research on the one hand, and biophilosophy and the
human condition on the other.

Topics are generally approached in terms of what is usually called
analytical philosophy of science, especially in its post-Kuhnian development. Since biophilosophy is not exhausted by this way of
thinking, topics are also illuminated from other philosophical perspectives such as oriental philosophies and points of view representative of our counterculture. Life is a phenomenon so vast and profound that strict adherence to one way of thinking, such as the
analytical philosophy of science, cannot do it justice. Since openness is a basic feature of life, philosophizing in a closed system can
only lead to an utterly fragmented view of life with the too well
known destructive consequences for society and the environment.

Since this is a book on biophilosophy, the topics are approached
in a general and abstract fashion. Many biological examples are,

however, presented in order to relate the general statements to concrete instances (e.g., the discussion on causality is illustrated in terms of the cancer problem and other examples). In addition, one chapter on comparative plant morphology has been included. Its intent is the reciprocal illumination of comparative plant morphology and biophilosophy. Thus, biophilosophy is exemplified through statements of comparative plant morphology; biophilosophical postulates are evaluated from the point of view of comparative plant morphology; and the philosophical basis of comparative plant morphology is explicated and scrutinized. Since progress and/or a better understanding of comparative plant morphology or any other biological discipline are intimately related to their philosophical foundation, case studies dealing with the reciprocal illumination of biological disciplines and philosophy are important for the philosopher and biologist. I have chosen comparative plant morphology as the subject matter for such a case study because I have carried out empirical, theoretical, and biophilosophical research in this discipline for over 20 years (e.g., Sattler 1966, 1973, 1974a,b,c, 1977, 1978a, 1982, 1984).

Diagrams have been included in a number of instances to facilitate understanding. A summary has been written for each chapter to provide an overview. Furthermore, an extensive bibliography of about 500 references has been compiled. These references range from technical to popular literature and present diverse perspectives of biophilosophy, biology, and life. They will enable the reader to pursue many of the topics dealt with or alluded to in greater depth and detail than is possible in this introductory exploration.

*Rolf Sattler*

# Acknowledgements

I wrote the first draft of this book during my last sabbatical leave in South-East Asia where I was collecting plants for my research in morphology. Thus, first of all, I am very grateful to my hosts: the late Dr. Kam Yee-Kiew at the University of Malaya in Penang, and Mrs. Thippavan Scott at Kratomtip on Phuket island in southern Thailand. Many weaknesses of the first draft were rectified as a result of the kind criticism by many students and the following colleagues and friends: Drs. Charles and Joanna Adams (Introduction), Prof. Masaaki Asai (Chap. 4), Dr. Réjane Bernier (Chaps. 7, 9, 10), Dr. Michael Bradie (Chap. 1), Mr. Sylvain Bournival (Chaps. 1 to 10), Ms. Claire Cooney-Sovetts (Chaps. 1-10), Dr. John Cornell (Chap. 7), Mr. Jean Caumartin (Chap. 1), Mr. Anthony Ferguson (Chaps. 1, 2, 8), Dr. Eduard O. Guerrant, Jr. (Chap. 1), Prof. Rainer Hertel (Chaps. 1-4), Dr. David L. Hull (Chap. 1), Prof. Aristid Lindenmayer (Chaps. 1, 9, 10), Dr. Robert H. Peters (Chaps. 1-10), Ms. Ghislaine du Planty (Chaps. 1-10), Mr. Yves Prairie (Chaps. 1 to 10), Dr. Rolf Rutishauser (Chaps. 1-10), Dr. Kurt Sittmann (Chap. 8), and Dr. Gunther Stent (Chaps. 1-4). I am also grateful to Ms. Celina Dolan and Ms. Joanne Smith for typing the manuscript, to Ms. Elspeth Angus for typing the index, to Ms. Doris Luckert for bibliographical assistance, and to Ms. Claire Cooney-Sovetts and Mr. Guy L'Heureux for the preparation and photography of a macerated maple leaf whose netted venation system served as the basis for the cover design. Finally, I want to thank the staff of Springer-Verlag, especially Dr. Dieter Czeschlik, Dr. Guido N. Forbath, Ms. Karin Gödel, Miss Ingrid Samide, Mr. Werner Eisenschink, and the Brühlsche Universitätsdruckerei for the highly competent, efficient, and rapid production of this book.

# Contents

XIV

5v (

XVI

# Introduction

**On the Importance of Biophilosophy.** Many biologists as well as non biologists tend to shy away from philosophy. A gut feeling or common sense makes them suspect that philosophy quite often loses close contact with real life, thus drifting into highly abstract constructions that bear little resemblance to nature. I think that this suspicion is justified to a great extent and for that reason I consider it desirable to eliminate or transcend philosophy. The crucial question is whether this is possible as far as science is concerned.

A number of scientists are optimistic in this respect. They point to history showing that certain problems that used to be endlessly debated by philosophers have become amenable to science. Thus, Wickler (1972), for example, elaborated on a "biology of the ten commandments." Other biologists have been debating for some time how evolutionary biology might provide a scientific basis for ethics. More recently, sociobiologists have attempted to establish a whole new discipline as a scientific basis for the behavior of animals and humankind (Wilson 1975, 1978; Barash 1977, 1979; Lumsden and Wilson 1981, 1983). Riedl (1980) proposed solutions to age-old philosophical problems in terms of an evolutionary epistemology, i.e., theory of knowledge (see Chap. 8). Accordingly, the philosophical foundations of science are said to be made scientific.

I think that these and similar attempts contribute to the interaction of science and philosophy including biophilosophy. However, none of them completely eliminates the philosophical basis of science and biology. Biology remains rooted in philosophical assumptions.

The borderline between philosophy and science may be fluid and even nonexistent so that philosophy of science and science of science intergrade. To some extent it depends on the definition of science whether a certain approach is still considered scientific or already philosophical. Hence, in certain cases it may be only a matter of definition whether an assumption is considered scientific or philosophical. In any case it remains an assumption that influences research.

2

In this book, I shall examine the most general and fundamental assumptions underlying biological research and its conclusions. Whether some of these assumptions are labeled philosophical or theoretical is of no absolute importance to me. I am more concerned with the consequences these assumptions may have for biology, our personal existence and society. I want to  show that they may fundamentally influence the quality of research as well as living. In fact our well-being and our survival may depend on at least some of them.

As a means of introduction to the foundations of biology, I shall now present three examples of assumptions that are often made in biological research. The first example is thinking in terms of a common factor which means that the same phenomenon must be caused by a common factor. Thus, alcohol consumption is seen as the causative common factor of drunkenness. The biophilosophical questions that confront us here are the following: How valid is thinking in terms of common factors? Can we take it for granted that this kind of thinking provides satisfactory answers in all cases? If not in all cases, in which cases is it appropriate and in which cases is it useless to search for common factors? Or is it generally misguided to hope that the idea of a common factor may provide explanations? To answer such questions we have to analyze the assumptions underlying the notion of a common factor. I shall briefly refer to only one of them that is not seldom implied in biological research. It is the assumption that the common factor has an inherent property or properties that cause a certain effect. When the common factor is a substance, that substance is thought to be endowed with a property or properties leading to a certain effect. Thus alcohol through the way it is constituted chemically is seen as the cause of drunkenness. In other words: the chemical constitution of alcohol unavoidably leads to drunkenness.

I think that this assumption is no longer tenable in the light of modern scientific and philosophical insights. One could go as far as to assert the opposite of the above postulate: a substance has no inherent nature that necessarily leads to a certain effect or effects. In short: a substance has no effect(s). Hence, a specific substance cannot be considered as a common factor for a certain effect. Why not? Because a certain effect is not fully determined by one or even a few substances (factors). It is determined by a substance (factor) or substances (factors) in a certain context or environment. Hence, it is the factor(s) as well as the context that determine the effect. The cause of the effect does not reside in the substance (factor) alone, but in the whole. Therefore, the search for substances (factors) that determine certain effects is questionable [for a detailed and rigorous analysis see, e.g., Weiss (1973), pp. 99-106].

Unfortunately, a great deal of biological research still seems to be aimed at discovering ultimate causes in terms of substances. Many publications have titles such as "the effect of (a specific chemical substance) ..." Such titles may suggest a belief in the essential nature of a substance that independently of its context causes the effect. It seems, however, that many authors are more or less aware of the importance of the context. In that case, the question arises: what is the general value of demonstrating the "effect" of a substance in one particular context? Generality, the principal aim of science (see Chap. 1), seems to be defeated in such approaches (which, of course, does not exclude their usefulness for specific purposes).

Applied sciences, such as pharmacology, also appear to suffer from thinking in terms of common factors (see, e.g., Weiss 1973). Instead of investigating the whole context, research is often focused on the "effect" of a particular substance as if it would cure certain diseases. Yet examples abound that demonstrate the importance of context. Depending on the physical and mental state of a person, LSD may have strikingly different "effects": it may "cause" heaven or hell. Evidently that which "causes" heaven or hell is not LSD alone, but the whole context which includes the person who is taking LSD and his or her environment.

These biophilosophical considerations have important consequences for biological research. They emphasize the importance of systems thinking that analyzes the whole (see, e.g., Bertalanffy 1967, 1968; Laszlo 1972b; Gray and Rizzo 1973; Gorelik 1975; Wuketits 1978, 1983). They suggest the development of multivariate models of analysis and thinking in terms of multidimensional spaces. Fortunately, more and more biologists are employing such methodology. As a consequence they are not only better equipped to deal with the integrated wholeness of organisms and ecosystems, but they are also able to show in which particular instances the simple approach in terms of common factors may be useful. Whenever certain variables of the whole context are constant or nearly constant, an explanation in terms of common factors may be warranted. It works in those particular instances, but for a different reason than assumed by essentialists who attribute an inherent power to substances or factors.

The example on thinking in terms of common factors shows that biophilosophical analysis is important in several respects: It places research into broader perspectives and it provides guidelines for specific research projects. Hence, it is not only important for the generalist and philosopher, but it has also crucial significance for the scientist in the laboratory or the field. Thus, biophilosophy is not necessarily armchair philosophy of an esoteric nature as many biologists tend to think, but is of fundamental practical importance because it is at the roots of all research.

As a second example of an assumption underlying biological research I want to point out that philosophical assumptions occur not only in the answers given to problems of research, but also in the questions asked in the laboratory or the field. Thus, most biologists quite often ask the following question when they encounter an unidentified plant or animal: Does individual X belong to species A or B? This kind of question may imply the belief that the individual X must be either species A or B, i.e., this question may be based on an *either-or* philosophy which implies that nature is discontinuous in that case. It excludes the possibility of continuous variation, in the above example at the species level. Consequently the idea of intermediates between species is not taken into consideration and accordingly not reflected in the question asked.

Research in systematics has shown, however, that boundaries between species are not always sharp. There are cases of intermediates between so-called species. At least in these cases the assumption of an *either-or* philosophy at the species level is not justified. Hence, it is not appropriate to ask an *either-or* question. Such a question would constitute a pseudo-question in that case.

Pseudo-questions are based on pseudo-problems, such as finding discontinuities in a continuum. They exist not only in systematics but probably in all areas of biology (see, e.g., Wuketits 1978, p. 38; Sattler 1978a). It is important to detect pseudo-questions and pseudo-problems because they are a hindrance to research. Biophilosophical analysis may help to separate pseudo-questions from useful questions, although there are cases in which it may be debatable whether a question is appropriate or not.

Practicing biologists often take it for granted that if an answer cannot be given to a question we just have to work harder and eventually the answer will appear. Such optimism is not warranted in the case of pseudo-questions. In fact, it will lead to a waste of time, energy, and money. So, again, we can see that biophilosophical analyses may have very concrete, practical, and "down-to-earth" implications.

Not seldom it is said that common sense can protect us from pseudo-problems and other nonsense. Although this may be true in some situations, it is not generally the case. Common sense is not infallible. It may reflect cultural, philosophical, scientific and religious prejudices and thus prevent us from breaking through traditionally biased ways of thinking and perception. *Either-or* questions and thinking in terms of common factors are examples.

The third example that I shall present here in support of the importance of biophilosophy has a rather metaphorical meaning. It is the question: do we have a soul? An answer given by a famous surgeon was the following: I have dissected many corpses, but I never found a soul. One might ask, did he really want to find a soul? Others who want to find the soul, experience it. Does this mean that we find what we want to find? In more general terms: do our preconceptions and our expectations influence our perception of the world? A whole range of answers has been given to this question from affirmation to negation (see Chap. 3). Although it may be impossible to reach agreement on this issue, it is important for biologists to reflect on it because it touches the core of all empirical science, which is based on facts. For if even the facts of science are controversial and relative, what can we expect from generalizations based on them? What then is the strength and value of science? Because of the fundamental importance of this issue, I have devoted a whole chapter to facts [for an excellent and penetrating discussion see also Woodger (1967), pp. 15–20].

My conclusions on the importance of biophilosophy are as follows. All biological statements and questions have theoretical and philosophical foundations. We can grasp the full significance of biological statements and questions only to the extent that we are aware of their foundations. Hence, biophilosophy concerned with foundations is of paramount importance to biology.

One can, of course, carry out good biological research without a knowledge of foundations. That does not mean that such research lacks foundations. It simply means that the researcher is not aware of them. To the extent that such subconsciously implied foundations are appropriate, the conducted research may be very successful. However, the scope of this somewhat blind approach is limited. Major breakthroughs of a revolutionary nature that involve a change of philosophical foundations are unlikely (see also Mohr 1977, p. 29).

In contrast to research that is oblivious of its foundations, biophilosophical awareness provides a more comprehensive understanding of biology and a prerequisite for innovation at a more basic level than that of "normal science" (see Chap. 1). Furthermore, biophilosophical awareness has existential and social dimensions. The most basic questions we ask as social human beings cannot be completely answered at the level of scientific statements. They refer particularly to the foundations of these statements. They lead us to the bottom of things which upon close inspection will disclose itself as bottomlessness ... no-thingness ... wholeness and unity (see, e.g., Izutsu 1971, 1974; Rorty 1979).

**On Definition of Biophilosophy**. The definition of biophilosophy can be approached in two ways: either through definition of "philosophy" or "philosophy of science." In terms of the first approach, the definitional problem presents itself as follows. Biophilosophy, which is the philosophy of biology, will be defined if we succeed in defining "philosophy" and "biology."

Biology can be defined as "science of life" which in turn requires a definition of 'science' (see Chap. 1) and 'life' (see Chap. 10). To avoid the difficulties involved in defining 'life,' biology could be defined as 'science of living organisms,' or, more broadly, 'science of living systems.' The latter definition includes all levels of organization from the molecular and cellular to the ecosystems level.

Defining philosophy is a hopeless undertaking because each original philosopher (or school of philosophy) has his or her own definition, if an attempt at definition is made at all. Many philosophers recognize this difficulty and refrain from futile formulations.

The term 'philosophy' is derived from two Greek roots: *philos* meaning love, and *sophia* meaning wisdom. Thus, the literal meaning of philosophy is 'love of wisdom.' Note that it is not 'possession of wisdom.' What we want to possess we shall lose. So if we want to possess wisdom by defining it, it shall escape us and we shall be left with rigid dogmatism or even fanaticism. It is a pity that many philosophers have forgotten the original meaning of philosophy when they force their philosophical systems upon others in absolute terms. And it is also most regrettable that many so-called lovers confuse love with possession and therefore fail to love profoundly (see, e.g., Fromm 1956). Possession is fixation of one state of affairs. The biologist knows very well that fixation means killing and death. Hence, in the long run possessiveness cannot but lead to destruction, be it of wisdom or love or whatever.

With such vague statements the scientist and philosopher who are striving for exactness will be utterly displeased. I shall therefore turn to the other approach of defining biophilosophy. Instead of beginning with philosophy in general, I shall restrict myself to a subdiscipline of philosophy. Philosophy has many subdisciplines such as logic, metaphysics, ethics, philosophy of language, epistemology (philosophy of knowledge), philosophy of art, philosophy of science, etc. Philosophy of science is the subdiscipline of interest here. If we can give a definition of philosophy of science, then philosophy of biology can be defined as that particular philosophy of science that deals with biological science. Hence biophilosophy is determined as a subdiscipline of philosophy of science.

So, how can philosophy of science be defined? It can be defined in terms of its task which is to analyze scientific statements including the reasoning through which they have become established. Nagel (1961, p. 14) distinguishes three broad areas of such analysis: (1) logical patterns of explanation, (2) construction of scientific concepts, (3) validation of scientific conclusions.

On the basis of this definition of philosophy of science, biophilosophy can be defined as the analysis of biological statements including the reasoning through which they have become established. The same three areas of analysis apply here with respect to biological statements.

This definition is useful, but it has shortcomings. The exactness it provides is at the expense of richness, i.e., the definition excludes interesting biophilosophical topics such as bioethics, the relation of biology to art and religion, culture and society, or simply the relation of biology to real living as a profound experience.

Considering the insurmountable difficulties of the first approach to a definition of biophilosophy in terms of "philosophy," and keeping in mind the shortcomings of the second approach in terms of "philosophy of science," I shall now follow a third approach which I think is more "down-to-earth" and not limited from the start. It is the approach followed in this book. Realizing that biology is the *science* of living systems, we are immediately led to the question of what is science and what is the aim of science. Obviously, scientific propositions such as theories and hypotheses play an important role in science. This poses the question: is it possible to demonstrate the truth or falsity of theories and hypotheses? If not, can we show which of two rival theories is the better one? In other words: is scientific progress possible? (Chap. 1). Laws, explanation, and prediction are also of major concern here. So we have to ask what they are, how they are obtained and founded. Is there lawfulness in nature or do we project our laws into nature? (Chap. 2). Since theories, laws, explanations, and predictions relate to facts, the next question is: what are facts? Are they hard and solid or totally or partially our own creations? (Chap 3). Facts as well as generalizations, such as theories and laws, are formulated in terms of a language that utilizes concepts and is based on a syntax and logic. Hence, the next question concerns the meaning of concepts (Chap. 4). Quite often biologists enquire about the cause(s) of biological phenomena. We have to ask: what is the meaning of this kind of question? Is it valid in all cases, in some cases, or not at all? (Chap. 6). Another general question often asked by biologists is that of purpose. What does that mean, and is there purpose in nature? (Chap. 7). Evolution and change are considered basic properties of living systems and evolutionary theory is usually accepted as the most comprehensive theory of biology. What are the foundations of evolutionary theory? Are they well established or are there still gaps and open problems? What is the relation between evolutionary thinking and epistemology (theory of knowledge)? (Chap. 8). And finally, the most fundamental and burning question of biology: what is life? (Chap. 9). Can we hope to find a complete and absolute answer as biologists or will our answer be limited because of inherent limitations of the scientific approach? (Chaps. 9, 10). If our scientific answer to life is limited, it cannot totally represent life as it is. So the question is: what is the relation between the science of life and actual living? Keeping in mind this question and feeling alive will anchor us in reality so that we shall not drift off into clouds of purely speculative systems whatever their seductive attractions may be.

# 1  Theories and Hypotheses

"Science *probes*; it does not prove" (Bateson 1979, p. 32)

"Theoretical statements, it is clear, cannot be verified because we can never know whether they are true. All we can do is to go on testing their consequences until an observation record turns up which contradicts them. Then we have the choice of two courses: we can say that the theoretical statement is false and reject it; or we can assume that we have been mistaken in our observations and retain the theoretical statements" (Woodger 1952, p. 57)

## 1.1  Introduction

In this chapter I shall deal with scientific generalizations, specifically theories and hypotheses. Other forms of scientific generalization shall also be mentioned. I shall ask and try to answer fundamental and crucial questions such as the following: What is the aim of science? How do we gain scientific knowledge? How do we validate scientific theories? Is there certainty in science either in the form of verification (proof) or at least in the form of falsification (disproof)? If not, is progress in science possible and what then can be a realistic reconstruction of scientific methodology? In short: What is the scientific status of theories and hypotheses? Are they dependent only on logic and empirical data, or are they in addition influenced by psychological, social, and cultural factors?

## 1.2  The Aim of Science

The aim of science is to gain knowledge of the world, in the case of biology knowledge of the living world. Knowledge has many forms. Weigel and Madden (1961) distinguish empirical perception through categories, metaphysical intuition, reason, faith, awareness or pure consciousness, and mystical experience. Pure consciousness and mystical experience go beyond knowledge inasmuch as they are a state of being. Science is only concerned with scientific knowledge which is expressed in singular and general propositions. Singular propositions are also called facts, whereas general propositions are referred to as hypotheses, models, rules, laws, and theories. As we shall see later on, the difference between singular and general propositions is not absolute or fundamental, but rather a difference of degree.

Theories are the most comprehensive and the best confirmed general propositions of science. They are the ultimate aim of science; the more

10

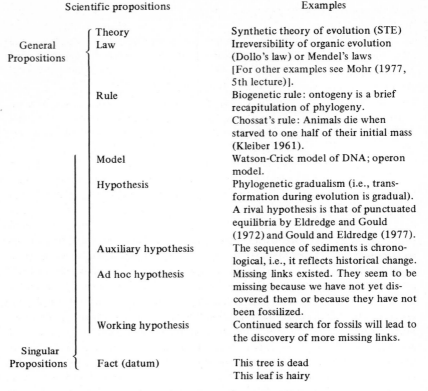

Scientific propositions — Examples

|  |  |  |
|---|---|---|
| General Propositions | Theory | Synthetic theory of evolution (STE) |
|  | Law | Irreversibility of organic evolution (Dollo's law) or Mendel's laws [For other examples see Mohr (1977, 5th lecture)]. |
|  | Rule | Biogenetic rule: ontogeny is a brief recapitulation of phylogeny. Chossat's rule: Animals die when starved to one half of their initial mass (Kleiber 1961). |
|  | Model | Watson-Crick model of DNA; operon model. |
|  | Hypothesis | Phylogenetic gradualism (i.e., transformation during evolution is gradual). A rival hypothesis is that of punctuated equilibria by Eldredge and Gould (1972) and Gould and Eldredge (1977). |
|  | Auxiliary hypothesis | The sequence of sediments is chronological, i.e., it reflects historical change. |
|  | Ad hoc hypothesis | Missing links existed. They seem to be missing because we have not yet discovered them or because they have not been fossilized. |
|  | Working hypothesis | Continued search for fossils will lead to the discovery of more missing links. |
| Singular Propositions | Fact (datum) | This tree is dead This leaf is hairy |

**Fig. 1.1.** List of different kinds of scientific propositions with examples [For more examples see Miller (1978)]

general and the more confirmed they are, the better. Thus, the ultimate aim of biology is a comprehensive and well-confirmed theory of life. We are still far away from such a theory, but we have biological theories of a general nature that have been confirmed to a certain extent. An example of such theories is the Synthetic Theory of Evolution (Fig. 1.1). (For a discussion of the semantic view of theories see Sect. 1.5).

Laws are also well-confirmed general propositions, yet they are less comprehensive than theories. Thus, "any (well-confirmed) general proposition within the framework of a theory may be called a law" (Mohr 1977, p. 48). One should add that the formulation of laws may precede that of the framework of the theory to which Mohr referred. Normally, two kinds of laws are distinguished: deterministic laws and probabilistic laws (see Chap. 2). It is probably fair to say that most of the biological laws are probabilistic. Many of them have numerous exceptions and thus are more appropriately called rules. It is disappointing to many biologists who search for lawfulness that so often general propositions which have been thought of as laws degenerate into rules as more and more exceptions to the law become known (see, e.g., Kochanski 1973, p. 42). In some instances, the rules may also dissolve or become questionable. An example is Haeckel's postulate that ontogeny is a brief recapitulation of phylogeny. At first proposed as a law, it was later on considered a rule (see Fig. 1.1) and has been questioned even as a rule (see, e.g., Dullemeijer 1974, p. 218; Gould 1977; Voorzanger and van der Steen 1982).

Hypotheses are propositions that in contrast to theories and laws have not yet reached a high degree of confirmation. Since confirmation is a matter of degree, no clear-cut distinction can be made between the general hypotheses on the one hand and theories and laws on the other hand. As far as degree of generality is concerned, hypotheses may range from very general and comprehensive propositions to rather specific, even singular propositions. In the latter case, hypotheses may coincide with facts, since facts contain a hypothetical element, and for that reason have also been termed low-level hypotheses (Lakatos 1968, 1978). Thus, very widely defined, hypotheses range from factual statements (singular propositions) to very general propositions of differing degrees of confirmation which in the extreme case border on theories. One might even include theories as extreme cases of confirmation among hypotheses, although this is not customary. In this sense, one would state that all scientific knowledge is hypothetical ranging from very specific propositions to very general ones, and from very questionable propositions to highly confirmed ones, the latter being customarily referred to as theories, or also as hypotheses if one opts to choose an unusually wide definition of the term. For my discussion of scientific methodology, this widest definition of hypotheses will be useful. To avoid confusion, I shall always state when I use the term hypothesis in this widest sense. Otherwise the term hypothesis shall be used in the ordinary sense as defined above, namely as a general proposition that excludes theories and laws at one end of the continuum, and facts at the other end. In this sense, several special kinds of hypotheses can be distinguished: auxiliary hypotheses, ad hoc hypotheses, and working hypotheses (Fig. 1.1).

Auxiliary hypotheses are additional assumptions that are necessary for the testing of any hypothesis or theory. No hypothesis or theory can be tested in isolation. Auxiliary hypotheses are thus unavoidable. What may constitute an auxiliary hypothesis in one context, may be the main hypothesis to be tested in another context. Hence, it depends only on the context whether a hypothesis functions as the main hypothesis or as an auxiliary assumption (auxiliary hypothesis).

Ad hoc hypotheses are additional assumptions whose function is to save and to protect the main hypothesis or theory in the case of contradictory facts, without leading to additional testable consequences. Thus, if facts are discovered that are in disagreement with the main hypothesis or theory, an ad hoc hypothesis can be invented in order to remove the contradiction. Since, in principle, such ad hoc hypotheses can render hypotheses and theories immune against observable facts, they violate the principle of empirical testability of scientific propositions. In other words: they violate the basic postulate of any science that it should be in agreement with empirical evidence, i.e., observable facts. For this reason, ad hoc hypotheses, strictly speaking, are not admissible in science. Unfortunately, as, for example, Hempel (1966) has pointed out, it is not easy to develop stringent criteria by means of which clear-cut distinctions can be made in all cases between ad hoc hypotheses and auxiliary hypotheses. Furthermore, what may be considered an ad hoc hypothesis at one time of scientific investigation may become acceptable as an auxiliary assumption at a later time (for examples of ad hoc hypotheses see Fig. 1.1 and Chap. 5).

Working hypotheses are highly hypothetical guidelines (heuristic devices) for the conduct of research (see Fig. 1.1). Even if they were wrong, they would fulfill an important function, because they provide the scientist with a framework that allows him to ask questions which in turn lead him to the design of experiments or the observation of new facts. Thus, working hypotheses propel the scientific enterprise and in this fashion may contribute to the formulation of new hypotheses, models, laws, and theories.

Models – as they are used in scientific methodology – have been defined in many different ways. A large number of special kinds of models have been distinguished and various classifications of models exist (see, e.g., Beckner 1959; Bunge 1973; Bertalanffy 1975, Chap. 8). For this reason it is difficult to give a general definition of the term 'model' that would be acceptable to all biologists and philosophers of science. I would tend to state, however, that models are scientific propositions that represent nature in an (over)-simplified fashion. Because of their partial agreement with nature, they may be predictive to a certain extent. According to Levins (1968, p. 6) "a model is built by a process of abstraction which defines a set of sufficient parameters on the level of study, a process of simplification which is intended to leave intact the essential aspects of reality while removing distracting elements, and by the addition of patently unreal assumptions which are needed to facilitate study."

The notion of the model overlaps considerably with the notions of hypothesis, rule, law, and theory. In fact, it has become customary for many biologists to refer to biological generalizations as models instead of hypotheses, rules, laws, or theories. That does not mean that no distinctions can be made between all those notions, but it shows that the concept of model is used in a very general way. One could go as far as to say that model building and model testing have become the main occupation of many biologists. (For an example see Fig. 1.1).

*Facts* (also referred to as data) are singular propositions that constitute the empirical basis of science (Fig. 1.1). Sometimes even well-confirmed generalizations are considered facts. For example, it is stated that evolution is a fact. I do not use the term fact in this sense.

Since facts are of utmost importance in science, a whole chapter is devoted to facts (Chap. 3).


## 1.3 Scientific Methodology or How We Gain Scientific Knowledge

### 1.3.1 General Considerations

The central and crucial question of scientific methodology is: how do we gain scientific knowledge? One facile answer to this question (which, for example, is often given in philosophical introductions to general biology texts) is the following: we gain scientific knowledge through the application of *the* scientific method, thus assuming that there is one single method that characterizes or even defines science. A study of the literature of philosophy of science quickly reveals, however, that there is no agreement at all whether scientific knowledge is gained by just one method called *the* scientific method. There are authors who claim that scientists use a variety of methods which cannot be subsumed under the heading of one scientific method (e.g., Hanson 1958) and there are philosophers and historians of science who are altogether "against method" (Feyerabend 1975, 1981a,b, 1982). Thus, among present day philosophers of science a whole range of opinions and attitudes exists from the belief in one scientific method to scepticism toward or denial of a method or methods that define science.

Regardless of the disagreement and controversy over the scientific method, it is generally acknowledged that scientific knowledge results from the conscious or subconscious application of some methodology. In fact, methodology is thought to be of paramount importance because it influences the knowledge. It determines the strength and the limitations of the knowledge. Therefore, to understand knowledge, in our case scientific knowledge, we have to be aware of the methodology through which the knowledge was obtained.

To demonstrate the importance of methodology to the biologist, it might be useful to begin with a brief discussion of practical laboratory methodology. Such methodology is not of direct concern to the philosopher of science whose interest lies in methods of reasoning. Nonetheless, an analysis of laboratory methodology also demonstrates convincingly that the results obtained are influenced by the methodology employed. Again, this methodology may be used consciously or subconsciously. In any case, it determines the strength and the limitations of the results.

I take as an example laboratory methodology that is used for the preparation of a microscope slide of a cross-section through a leaf. It would be naive to assume that looking at this slide through the microscope would reveal to us the structure of the leaf, as it exists in nature, in a transverse view. What we actually see is how the leaf has been transformed by our laboratory methodology.

Two sets of methods may be distinguished: preparation methods and methods of microscopy. The preparation methods include killing and fixation of the leaf, its dehydration and embedding, sectioning, removal of the embedding substance, staining of the cross-section of the leaf, and mounting on a slide in a preserving substance under a coverslip. All of these methods influence the natural structure of the leaf to some extent. Unless we are aware of those methods and how they affect the natural structure, we are unable to distinguish natural structure from artifacts due to the preparation methods.

In addition to the preparation methods, our methods of microscopy influence the picture that we observe. For example, the type and intensity of the light and the kind of magnification of the optical system affect the observation we make of our slide. Finally, we may add that the methodology of our visual perception profoundly influences our picture. Specifically the construction of our eyes and our brain is responsible for our perception, but we may look upon the operation of our whole physico-mental organization as a methodology that influences the picture we obtain of that cross-section of a leaf (see, e.g., De Duve 1984, p. 17).

In this connection, the doctrine of *Umwelt* by Jacob von Uexküll (1920, 1957) may be mentioned. According to this doctrine each animal species perceives the world differently because of its different organization which implies a different methodology of perception. Hence, there are as many *Umwelten* (ambient worlds) as there are animal species. The human *Umwelt* is one among them. It would be naive to assume that we perceive the world as it is. In fact, we know that many animal species perceive aspects of the world that are totally beyond human perception. For example, bees may see patterns of ultraviolet in flowers that we cannot perceive at all. And who of us can imagine the *Umwelt* of a dog with its richness and intensity of smells! Science has provided us with the theoretical knowledge of the variety of perception in other species. But the direct experience of other species has not been revealed by it.

Returning to our perception of the prepared cross-section of a leaf, we can state now that it depends on the natural structure of the leaf as well as our laboratory methodology and our methodology of perception which is determined by our organization. Hence, a knowledge of the cross-section of a leaf requires understanding of methodology that necessitates a knowledge of ourselves. We cannot exclude the observer if we want to understand the observed: both form a unity (e.g., Bateson 1979).

Perhaps it is easier now to imagine the problems involved in the knowledge of the cross-section of a leaf. If we communicate this knowledge in an objective and conceptualized manner in the form of singular propositions (facts), we have to take into consideration the additional problem of the relation between this knowledge and its linguistic representation. This problem will be briefly discussed in Chap. 3. What matters at this point is that we realize that so far, in discussing our perception of the cross-section of a leaf or any other feature of the world, we have been dealing only with facts which may be represented by singular propositions. Although facts are the basis of science and thus of utmost importance, they do not yet constitute science. Science is not a mere collection of facts. As I have stated already, science is theoretical, i.e., its aim is theories. In order to obtain theories, a methodology is required that for the sake of communication I call "scientific methodology." I shall turn now to a discussion of this scientific methodology that belongs to the domain of philosophy of science and biophilosophy, i.e., the main topic of this first chapter.

Since it is not easy to be fully aware of the scientific methodology one employs as a scientist, it is necessary to reconstruct it (e.g., Tweney et al. 1981). Reconstruction is a difficult task and has led to controversies that have been briefly alluded to above when I referred to the disagreements about the scientific method. In the following, I shall begin with an illustration of simple – in my opinion oversimplified – schemes of the so-called scientific method and shall proceed gradually to the complex and complicated views of scientific methodology which – in my opinion – come closer to the process of science as it is practiced by scientists. I think that those latter views (such as, for example, the system model of scientific methodology) represent the best reconstruction of scientific methodology available today. However, the inclusion of simplistic reconstructions permits the discussion of several important issues of philosophy of science.

## 1.3.2 Induction and the Hypothetico-deductive Method

We grow up in a certain society in which we are more or less influenced by tradition with its scientific, philosophical, and religious roots. In other words:

16

we are formed within a certain cultural setting. If we become scientists we are in addition shaped by the scientific community, i.e., our teachers, the literature we read, and our daily discussions on scientific issues. In this context, we start to do scientific research and we continue it interacting with the scientific community and society in general. Thus, the first scientific problem we tackle and the first observation we make as well as all the following scientific problems and observations with which we are confronted as scientists occur in a cultural and scientific context (see, e.g., Jonas 1966). It is important that we remain aware of this. Some authors who discuss cuss scientific methodology treat it independently of this background as if one could make an observation or formulate a problem without being significantly influenced by the cultural and scientific context.

When we make an observation now and formulate it as a singular proposition, the question that arises immediately is: how can we arrive at a general proposition which is the aim of science? One answer to this question is the following: it is by means of induction (or inductive generalization) that we may proceed from a singular proposition to a general one. Induction in its simplest form can be easily illustrated by one of the famous examples used by philosophers of science. We observe one swan and note that it is white, and then we generalize inductively that all swans are white (see Fig. 1.2).

Fig. 1.2. Induction illustrated by a simple example

Inductive generalizations do not carry logical force, i.e., they are not arrived at by deductive logical reasoning. From the fact that one swan is white, it does not follow logically that all swans are white. Hence, inductive generalizations cannot be proven. Even if many additional observations would support the inductive generalization, one could not be sure that future observations would also be in agreement with it (see below). Thus, inductive generalizations are constantly open to revision. They are not established truths. Even if they were correct, we would have no way of demonstrating (proving) it.

Although one can hardly deny that induction in the above and other more complex forms plays a role in science, it would be inappropriate to claim that it is the only method that generates general propositions (see, e.g., Cohen and Hesse 1980). It may not even be the most important method. As soon as the scientific problem reaches a certain level of complexity, intuition and deductive reasoning are required in order to arrive even at a tentative solution (e.g., Hempel 1966; Medawar 1969). The method that can be employed then is the 'hypothetico-deductive method.' It can be presented in a simplified diagrammatic form as follows (Fig. 1.3):

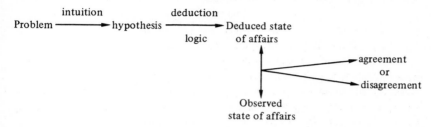

**Fig. 1.3.** Simplified representation of the hypothetico-deductive method. (Compare with Fig. 1.6, p. 28)

An example illustrating this method is the elucidation of the structure of benzene by Kekulé (Hempel 1966, p. 16). The problem was the structure of the benzene molecule. Observations of the reactions of benzene did not lead directly to any solution of the problem. What was needed was an idea, a "happy guess" (as Whewell, quoted by Hempel 1966, p. 15, put it) in order to arrive at a hypothesis which, at first, is just a conjecture. Such a conjecture is the result of intuition (which comprises a complex form of induction, according to some authors). So, the story goes, one evening Kekulé "was dozing in front of his fireplace. Gazing into the flames, he seemed to see atoms dancing in snakelike arrays. Suddenly, one of the snakes formed a ring seizing hold of its own tail. Kekulé awoke in a flash: he had hit upon the now famous and familiar idea of representing the molecular structure of benzene by a hexagonal ring" (Hempel 1966, p. 16). Now with a hypothesis in mind, Kekulé could make deductions from this hypothesis. He could say that if this hypothesis is correct, then it follows logically that this or that must be the case. Thus he could deduce certain states of affairs. He then could make observations of states of affairs that occurred naturally or were the outcome of an experiment. In principle, a comparison of the deduced states of affairs and the observed states of affairs yield either agreement or disagreement. In the case of Kekulé's hypothesis, the observations agreed with his deductions and therefore he was confident that he found the right solution to the problem. If the observed states of affairs would have contradicted the deduced ones, he probably would have looked for another hypothesis, which means that he would have needed another intuition.

18

In summary, we can state that three elements are of major importance in the hypothetico-deductive method, namely intuition, logic in the form of deductive reasoning, and empirical input in the form of facts. All of these must be seen in relation to the cultural setting whose influence will be discussed under the heading of the third postulate of validation and the general systems model of scientific methodology. [For a criticism of the hypothetico-deductive method see, e.g., Butts (1976)].

### 1.3.3 Validation of Hypotheses, or: Is Certainty Attainable?

We have already seen that inductive generalization does not lead to statements whose truth can be asserted. Does the same apply to statements arrived at by the hypothetico-deductive method? More specifically, does agreement of the deduced and observed state of affairs entail proof of the hypothesis under consideration? Or, vice versa, does disagreement mean disproof? In more general terms: is certainty of scientific knowledge possible? If not, what is the status of validation of general scientific propositions? In the following section I shall examine three postulates (or philosophies) with regard to the status of scientific knowledge. In this connection, I shall base my discussion primarily on the hypothetico-deductive method, and I shall employ the term 'hypothesis' in its widest sense so that it includes theories. The reader who does not approve of this definition of 'hypothesis' may replace 'hypothesis' by 'hypothesis including theory.'

#### 1.3.3.1 *The First Postulate of Validation*

This postulate can be formulated as follows: *Hypotheses may be proved.* In other words: the truth[1] of hypotheses may be asserted.

According to this postulate or philosophy (which is also referred to as justificationism) we may write the scheme of the hypothetico-deductive method as follows (Fig. 1.4):

---

[1] I use the terms 'truth,' 'proof,' 'disproof,' and 'certainty' in the strong sense, i.e., meaning absolute truth, final proof and disproof, and absolute certainty. It might be objected that this is contrary to common usage. I am not convinced of that. For example, when someone says "It is true that it rained on the other side of the mountain," we normally understand this statement in the absolute sense. If we were not (absolutely) certain, we would say: Probably (or most probably) it rained on the other side of the mountain. Or we would choose some other expression that indicates literally that there is at least some doubt whether it actually rained.

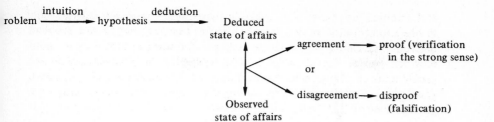

**Fig. 1.4.** The first postulate (philosophy) of validation: hypotheses may be proved (justificationism). The context which is indicated in Fig. 1.6 is omitted here

Justificationism is primarily connected with inductive methodology but it may also be referred to in the hypothetico-deductive context. To adherents of justificationism agreement between the deduced and observed states of affairs means proof of the hypothesis, whereas disagreement may be regarded as disproof. The emphasis is on proof because by proving one hypothesis after another one makes undeniable progress. Thus, science is seen as a linear progression from ignorance to knowledge, whereby knowledge is understood as proven knowledge. Certainty is possible according to this philosophy. As science progresses, uncertainty is gradually eliminated (e.g., Harris 1981).

This philosophy is basically optimistic as far as the acquisition of proven knowledge is concerned. It may require hard work, but by proving one hypothesis after another, we shall come closer and closer to the truth. Eventually we shall be in possession of the whole truth, and if for some reason this should not be possible, we can at least be assured that we shall approximate the whole truth more and more. Progress of science, according to this philosophy, can be compared to the construction of an edifice: one brick is put above another until finally the edifice is erected in its whole glory. Even if there are times in which progress is slow or lacking, we shall not despair because we know that we have at least an unshakable foundation of proven hypotheses on which we can stand with certainty when new proofs will enable us to continue the construction of the great edifice of truth. Scientific revolutions that may suddenly demolish our present knowledge are impossible according to this view because knowledge once proven is established forever. If one wants to use the term of scientific revolutions at all in this context, its only meaning can be a sudden and considerable increase of scientific knowledge which would amount to an enormous breakthrough of a fundamental nature.

The following botanical example shall illustrate the above postulate of validation. Botanists are interested in the causation of plant growth because growth is a fundamental process that leads to the whole diversity of plants.

It is known that to a great extent plant growth is due to cell enlargement. Hence the causation of cell enlargement has been considered as a problem of prime importance. If we want to tackle such a general problem, it is useful to work on one kind of organ in a convenient species. With regard to the problem of cell enlargement, one such object is the coleoptile of *Avena* (oats). A coleoptile is a cylindrical organ several centimeters long which encloses the young shoot of the *Avena* seedling. It has been known for a long time that the coleoptile grows mainly through cell elongation except at its tip. Therefore, with regard to the coleoptile our general problem of cell elongation can be stated as follows: what is the cause of cell elongation in the coleoptile? (For a critique of the notion of cause see Chap. 6). In order to propose a hypothesis (hypothetical solution) to this problem, one needs an idea, an intuition. Already in the first part of this century the following hypothesis was proposed. Auxines (i.e., growth hormones) are the cause of cell elongation. They are produced in the tip of the coleoptile and as they move downward they provoke cell elongation and thus growth of the coleoptile. On the basis of this hypothesis, several deductions (predictions) can be made. For example, if the tip of the young coleoptile is removed, the stump of the coleoptile should stop growing. This experiment was carried out long ago and it was observed that after removal of its tip, the coleoptile (gradually) ceased to grow. Thus, there is agreement between the deduced and the observed state of affairs and hence the hypothesis is proved, according to the first postulate (philosophy) of validation. After this success, other objects were investigated in a similar manner, additional agreements between predictions and observations were obtained, and thus growth of scientific knowledge occurred as outlined above.

The question that we have to ask now is whether this view of validation and the progress of science that it entails are tenable. My answer is no. I use the swan hypothesis as an example to illustrate my negative answer. Although this hypothesis is normally discussed in the context of induction (see above Fig. 1.2), it may also exemplify the hypothetico-deductive method. According to this hypothesis, which states that all swans are white, we can deduce (predict) that the next swan that we shall encounter will be white. If that swan turns out to be white, we have an agreement of deduced and observed states of affairs and hence, according to the first postulate of validation, proof of the white swan hypothesis. It is, however, evident that one successful prediction cannot constitute proof of that hypothesis. Even if we had succeeded in making a thousand (or more) correct predictions, we could not be assured that the next swan we observe will also be white. Thus, a most confidence inspiring success of predictions in the past does not at all guarantee future success. One might perhaps say that past success makes future success more likely. But even this is doubtful. The white swan hypothesis was accepted for a long time until one day black swans were eventually found. Nonetheless, an observer who limits his search to one area of the world in which only white swans occur would never fail to make successful predictions. He could die with the conviction that all swans are white (because he never found an exception) and yet he was wrong. However, as an analyst of scientific methodology he could come to the following realization: Although I have always encountered agreement between the deduced and observed state of affairs, my hypothesis has not been proven and never will be proven. I have to accept that all scientific knowledge is unproven knowledge. I have to realize that certainty is not attainable in science [for other examples see Polanyi (1964), p. 94].

Someone who clings to the ideal of certainty may not yet capitulate at this stage and argue as follows. Since the number of swans must be finite, we could hope to observe eventually all swans and thus prove the new hypothesis that all swans are either white or black. But how can we ever obtain certainty that we have observed all swans? It seems that certainty is really unattainable[2]. However, the scientist and philosopher who stubbornly hopes to reach certainty at least in some instances would continue to argue as, for example, Hartmann (1948) or Nachtigall (1972). These authors would admit that certainty is not possible in the swan hypothesis because it is a hypothesis that is validated by observations of naturally occurring events and things. They insist, however, that hypotheses validated by observations of experiments may be proven. Hence certainty is possible in experimental biology according to these authors. It is not difficult to demonstrate that their claim is untenable. If we consider our coleoptile hypothesis as an example of experimental biology, then we can see that even an experiment cannot constitute proof of an hypothesis. The fact that the coleoptile stops growing after removal of its tip is in agreement with our hypothesis, but this agreement could be due to different reasons: either because the hormone source has been removed as postulated by the hypothesis or because the system has been too much disturbed as a result of the injury of the coleoptile. Experiments entail interference with the object under consideration (even in a so-called controlled experiment). If we cut off the head of a human baby, it would also stop growing. How can we ever have absolute assurance that after the removal of the tip of the coleoptile it ceases to grow simply because the hormone source is lacking? It could stop growing for many other reasons. How could we ever devise experiments to test all of the other possible reasons? And if we could, how could we be certain that we have thought of all possible

---

[2] I am not claiming that the swan hypothesis is of great importance to biological theory, nor am I assuming that species are natural kinds (see Chap. 4).

reasons? So, how can we ever attain certainty in science, be it descriptive, comparative or experimental science?

The first section of the first chapter of Bateson's (1979) book on *Mind and Nature* is entitled: "Science never proves anything." It is probably fair to say that most present-day philosophers of science would agree. However, biologists appear to be more ambivalent in this respect. If one scans the biological literature, one encounters not seldom authors who try to verify or to prove a hypothesis. Unfortunately, in most cases they do not explicate what they mean. Literally, verification has the meaning of showing the truth. However, 'verification' has been used in a weaker sense as a synonym of 'confirmation' (Mainx 1967, p. 4). Hence, it is often not clear whether authors who use the term 'verification' subscribe to the first postulate of validation. Most authors who refer to 'proof' also fail to specify whether they use this term in the strong or weak sense, i.e., as absolute proof or as confirmation. When questioned, they often would point out that absolute proof does not exist in empirical science. This claim is, however, frequently contradicted by their behavior. They often dismiss criticism of current theories or hypotheses in the following way: Do not question this theory. It has been proven!

It is ironical that the first postulate of validation as I presented it is usually rejected, and yet many biologists behave as if it were correct, i.e., they seem to accept it subconsciously. Such subconscious acceptance does not render it less dangerous. Those who accept it – be it consciously or subconsciously – may be inclined to impose their views on others and if they happen to be in powerful positions they may repress those who refuse to accept their so-called truth. Unfortunately, history is full of such hubris, intolerance and repression (see, e.g., Bronowski 1973, Chap. 10). Even present-day science faculties of universities and research institutes are not always free of it. To some extent we are still living in the dark ages, one reason among others being that the first postulate of validation with its associated philosophical outlook and social repression has not yet been overcome completely. Nonetheless, we can say with Bronowski (1973, Chap. 10) that science and philosophy of science have shown us the way of liberation from the darkness of intolerance and repression based on supposed certainty to the recognition of uncertainty and tolerance. Let us hope that these intellectual insights may eventually penetrate deeply into our subconscious and thus provide our society with a more profoundly humane basis before it is be eradicated by those who think and act as if they knew the truth. Educators have a deep responsibility and a challenging task in this respect (see Weiss 1973, Appendix). They should beware, however, of conveying uncertainty and tolerance in a dogmatic and intolerant or even repressive form. If we cannot prove hypotheses, then we cannot prove that certainty is untainable. We can only make it evident.

Besides the social implications, there are also existential dimensions. One encounters often the belief that we need certainty in order to live a fulfilled life. Riedl (1980) and other evolutionary biologists claim that the urge for certainty has evolutionary roots and is therefore quite natural. I tend to think, however, that this urge results mainly from cultural conditioning, since in some cultures the preoccupation with certainty does not exist as in our Western culture. And even in our culture some individuals have freed themselves of this desire. The fact that many of us find it difficult or impossible to acknowledge uncertainty, does not "prove" that certainty is necessary for living. It simply shows that those clinging to the desire of certainty are using this ideal as crutches for living. As a result their natural ability to live freely has atrophied because of the constant culturally reinforced use of these crutches. In that situation it may indeed be unwise to throw away the artificial support from one day to the next because that might lead to a collapse in cases of extreme dependence and atrophy. But any person who has still some strength can learn to strengthen atrophied abilities again through increased use and thus will succeed to walk without crutches and finally may jump and dance. Moving and dancing in total spontaneity is the antidote to a worried life dominated by the urge for certainty. Many Eastern philosophies and ways of living are rooted in this spontaneity which is beyond the rigid quest for verbal truths and certainties (see, e.g., Lao Tsu's Tao Te Ching 1972; Watts 1951; Trungpa 1973, 1976; Rajneesh 1975a,b, 1978).

### 1.3.3.2 The Second Postulate of Validation

This postulate can be formulated as follows: *Hypotheses cannot be proved, but they may be disproved (falsified).* According to this postulate or philosophy (which is referred to as falsificationism) we may write the scheme of the hypothetico-deductive method as follows (Fig. 1.5):

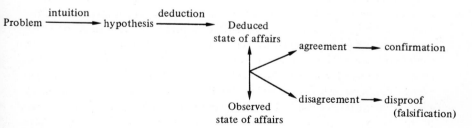

**Fig. 1.5.** The second postulate (philosophy) of validation: hypotheses cannot be proved, but they may be disproved (falsified) (falsificationism). (The context which is indicated in Fig. 1.6 is omitted here)

To adherents of this postulate agreement between deduced and observed states of affairs does not mean proof of the hypothesis, but simply confirmation. Confirmation (= corroboration) is not absolute, but a matter of degree (see, e.g., Carnap 1936/37; Hempel 1966; Martin 1967). Even the most highly confirmed hypothesis is shakable. Therefore, progress in science does not result from a gradual accumulation of proven hypotheses, but occurs through elimination of falsified hypotheses. It is assumed that as more and more hypotheses are falsified, we retain eventually those that are highly confirmed. The more often a hypothesis has survived attempts of falsification, the more confidence we have in it. But we shall never be committed to an hypothesis; according to this philosophy commitment to a hypothesis is a crime (Lakatos 1968, p. 150, 1976). Hence, falsificationism requires a critical attitude and considerable openness.

Scientific revolutions occur according to this philosophy in a rather dramatic way when a hypothesis or a whole paradigm is falsified and is replaced by an alternative. An example of the latter is the changeover from creationism to evolutionary thinking. If we compare again progress in science to the construction of a house, we can no longer rely on the foundations we have laid. At any time our whole construction may collapse and we may have to start all over again. The only certainty that the second postulate provides is that once we have falsified a hypothesis, it is eliminated from science forever and therefore we do not have to worry about it any more. We can then focus our attention on other hypotheses with the hope of demolishing them. Such demolition is nothing negative according to this philosophy. It is the best thing we can do. Criticism in the form of falsification is seen as the moving force of science.

An example that may illustrate the second postulate of validation is the falsification of the visual hypothesis of the homing behavior of silver salmon [for a review of the literature see Leggett (1977); Hasler and Scholz (1983)]. As is well known, the young silver salmon hatch in the streams of the Pacific northwest. They swim downstream to the ocean and finally a certain percentage of them returns to exactly the same stream in which they were born. The problem is: how do they find their way back? By sight or by odor? If we test the visual hypothesis, we can make the following deduction: blindfolded salmon will not find their way back. In this case the deduction (or prediction) will have to be formulated in statistical terms, but that does not change the basic principle of testing. We then observe that a certain percentage of individuals find their way back in spite of the blindfolding. As a result of this disagreement of deduced and observed states of affairs the visual hypothesis is disproved (falsified) according to the second postulate of validation. Hence, we can eliminate this hypothesis as false and can continue with the testing of other hypotheses. A test of the odor hypothesis is confirmed. Thus our temporary conclusion is that silver salmon find their way home by perceiving the distinctive odor of the river in which they hatched.

The question that we have to ask now is whether this view of validation and the progress of science is tenable. My answer is again no. Disproof of a hypothesis is not possible for at least two reasons:

(1) Facts, i.e., the observed state of affairs, cannot necessarily be considered to be true. As I have mentioned already, facts also contain a hypothetical element, although they may be much less hypothetical than the hypothesis to be tested. For that reason facts have also been termed low-level hypotheses or observational hypotheses. It is evident that one hypothesis, i.e., the high-level hypothesis to be tested, cannot be disproved by another hypothesis, i.e., the low-level hypothesis of the observed state of affairs. Hence disproof of the high-level hypothesis to be tested is not possible, since in the case of a contradiction between a high-level and a low-level hypothesis the falseness may reside in the low-level hypothesis (i.e., the fact). What does this mean with regard to the salmon example? It means that the observed state of affairs, namely that blindfolded salmon return to their home river, could be faulty. Although it might be extremely unlikely that it is faulty, one cannot absolutely guarantee that it is true. For this reason, the visual hypothesis cannot be eliminated once and for all. In order to illustrate the hypothetical element of facts, we may look at other examples of facts. "This tree is dead" is a fact whose hypothetical nature is easily recognized. How could we be absolutely sure that a tree that looks dead actually is dead? How hypothetical the fact is would depend, of course, on the specific case. In some instances it might indeed be rather questionable whether the tree is dead, whereas in other instances this statement may border on certainty. But we can not be certain even at the level of facts. Another example: "The shape of this leaf is elliptical." In this case it is even more difficult to imagine why this observation could be wrong. But maybe a closer inspection would reveal that it is not exactly elliptical. Maybe we got an altogether wrong impression. Illusion and mass hallucination have been reported in many instances (see, e.g., Gregory and Gombrich 1973). Thus, although it is difficult to imagine errors at the factual level, we cannot totally exclude the possibility that in a particular situation the fact may be faulty. Therefore, we have to reject the second postulate of validation [see also the quotation by Woodger (1952) at the beginning of this chapter].

(2) In the discussion of scientific methodology I have pretended so far – for the sake of simplicity – that a hypothesis can be tested separately from other hypotheses. This is not possible. A hypothesis is always tested in conjunction with auxiliary hypotheses. Therefore, in the case of a clash between the hypothesis to be tested and the fact, neither the hypothesis nor the fact need be faulty, but one of the auxiliary hypotheses could be wrong. If we

take a simplified case of only one auxiliary hypothesis, we arrive at the following scheme (Hempel 1966, p. 23), H being the hypothesis to be tested, A the auxiliary hypothesis, and e the fact:

If H and A, then e

*e is not the case*

H and A are not both true

With regard to the salmon example I might mention the following auxiliary hypotheses:

(a) The operation technique used for the experiment leads to complete blindness. Suppose that this auxiliary assumption is false, then the fact that the operated salmons (which we assumed were blindfolded) returned to their home river no longer contradicts the visual hypothesis.

(b) It is assumed that in the case of complete blindfolding the animal does not compensate for this visual loss by a regulatory mechanism which enables it to reach its goal by means different from those of the normal behavior. If this auxiliary hypothesis were wrong, then the success of the actually blindfolded animals again would not contradict the visual hypothesis because the animals adapted to the experimental condition.

(c) The percentage of operated animals that return home is statistically significant, i.e., cannot be explained by chance or other factors.

These examples may suffice to illustrate that disproof of a hypothesis is not possible because it is linked to auxiliary hypotheses. Very often it may be difficult or impossible to enumerate all auxiliary hypotheses because one may not be aware of them. In addition, the hypothesis to be tested is linked to the general background knowledge of the scientific discipline and even to metaphysical assumptions (Bunge 1977). Hence the scientific and philosophical background, which so often is taken for granted, may also be responsible for contradictions between facts and the hypothesis to be tested, i.e., the fault may reside in the generally accepted bulk of scientific and philosophical knowledge and the hypothesis to be tested may be correct. An extreme consequence of this is expressed in the Quine-Duhem conventionalist thesis according to which "any statement can be held to be true no matter what is observed, provided that adjustments are made elsewhere in the system" (Presley 1967).

In spite of the above arguments that one may level against the postulate of falsifications, there are still biologists who subscribe to it and go as far as to define a scientific hypothesis as a proposition that is falsifiable. The term 'falsification,' however, appears to be used differently by different authors, as I noted it for the term 'verification.' Thus, a similar ambiguity has been created as in the case of the first postulate of validation. It is possible that some authors use falsification in a weak sense meaning disconfirmation which is the appropriate conclusion in the case of a disagreement between the deduced and the observed states of affairs. These authors do not really subscribe to the second postulate of validation but represent the third postulate, although they insist on using the term falsification (in a weak sense).

Sir Karl Popper who has to be mentioned in a discussion of falsificationism has presented us with an interesting spectrum of ideas on falsification, refutation, and scientific methodology from his early to his later writings (see, e.g., Popper 1935, 1962, 1972, 1984). His former student Lakatos made a distinction between three versions of Popperian philosophy of science which he termed $Popper_0$, $Popper_1$, and $Popper_2$ [see Schilpp (1974) for Popper's response]. $Popper_0$ is dogmatic falsificationism, the philosophy of the second postulate of validation according to which falsification is used in its strong sense as equivalent to disproof. It is ironical that $Popper_0$ who has been invoked so often never existed. As Lakatos (1968, p. 152) pointed out, $Popper_0$ is a fiction of some philosophers of science, and we may add also a number of biologists. Already in 1935, Popper stated that "no conclusive disproof of a theory can ever be produced" [quoted from the English translation, 1959; see also Brown (1977, Chap. 5)].

$Popper_1$ is naive falsificationism. According to this philosophy, the high-level hypothesis must always be discarded in case of a clash with a low-level hypothesis (i.e., a fact). This philosophy is espoused in his works like *Conjectures and Refutations* (Popper 1962) in which progress of science is seen as elimination of hypotheses due to refutation. Famous $P_1$ slogans are: "Make sincere attempts to refute your theories", or: "A refutation is a victory" (quoted by Lakatos 1968, p. 162). Refutation is possible for $Popper_1$ because of his belief that a low-level hypothesis (i.e., a fact) is more trustworthy than a high-level hypothesis. Although this may often be the case, we have no assurance that this is always so. In fact, we have learned from historical experience that facts may be faulty (see Hempel 1966). Thus Lakatos' $Popper_1$ is indeed naive and no longer defensible.

$Popper_2$ concentrates on growth of scientific knowledge, not on refutation of hypotheses (Lakatos 1968, p. 162). The crucial question for $Popper_2$ is which hypothesis is the best among various rival hypotheses. The answer is the following (Lakatos 1968, p. 163): A hypothesis "is better than its rival a) if it has more empirical content, that is if it forbids more 'observable' states of affairs, and b) if some of this excess content is corroborated" (see also Lakatos 1976, 1978).

With $Popper_2$ and the still more recent versions of Popperian philosophy (e.g., Popper 1972, 1984) we have gone far beyond the outdated second postulate of validation to approach the third postulate which is a more appropriate philosophy of validation.

28

### 1.3.3.3 The Third Postulate of Validation

This postulate can be formulated as follows: *Hypotheses can be neither proved, nor disproved; they may only be either confirmed or disconfirmed.* On the basis of this postulate and keeping in mind the conclusions of the preceding sections, we arrive now at a much more realistic and comprehensive picture of scientific methodology (Fig. 1.6):

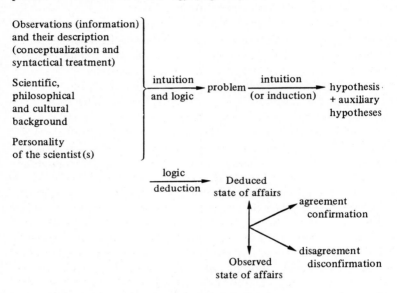

**Fig. 1.6.** A more realistic scheme of scientific methodology than the oversimplified versions of Figs. 1.3–1.5

In this reconstruction of scientific methodology it is indicated that the scientific enterprise does not occur in isolation from society with its scientific, philosophical, and cultural traditions and that the personality of the scientist also plays a role. Thus, problems arise not merely as a logical consequence of objective observations or prior scientific knowledge, but are also the results of creative insight and the whole context. As I have pointed out already, simple hypotheses may be generated by inductive generalization, whereas the invention of more complex hypotheses requires intuition. Regardless of whether hypotheses are simple or complex, they are tested in conjunction with auxiliary hypotheses. Any test can lead only to either confirmation or disconfirmation. Certainty no longer has a place in empirical science, neither in a positive sense (i.e., proof), nor in a negative sense (i.e., disproof). Thus, the third postulate differs drastically from the first two. If, however, proof in the first postulate and disproof in the second postulate are interpreted in a weak sense, both of these postulates merge with the third one. Thus, actually a continuum between the three postulates exists.

Since confirmation does not entail that the hypothesis is correct, and since disconfirmation does not mean that the hypothesis is false, the burning question is: how do we decide which hypothesis is acceptable and which is not? A hypothesis that has been confirmed in all tests could be provisonally accepted without any problem. Vice versa, a hypothesis that has been only disconfirmed can be provisionally rejected. However, if, as is usually the case, a hypothesis has been confirmed in some instances and disconfirmed in others, how are we to arrive at a rational and objective decision? How many disconfirmations are needed for a rejection of a hypothesis? There are no objective and rational criteria to answer this question satisfactorily. The history of science shows that the personality of the individual scientist, the scientific community and society at large may play an important role (see, e.g., Rose and Rose 1976a,b; Chant and Fauvel 1980; Coley and Hall 1980; Gould 1981; Thuillier 1981; Edge 1983; Lewontin et al. 1984). Thus, in addition to the objective inner-scientific (or internal) factors, psychological and social factors that may be referred to as outer-scientific (or external) decide whether a hypothesis is accepted or discarded. Scientific practice shows the interplay of these factors. In many cases individual scientists disagree. What is acceptable to one may be questionable or even totally unacceptable to another. The inner-scientific evidence may be the same to all of them, yet they weigh it differently: what to one of them may be an irrelevant anomaly, to another one may be an important contradiction which, in his opinion, justifies the rejection of the hypothesis. What is the greatest discovery to one of them may be utter nonsense to another one. However, there are hypotheses that have been accepted by a vast majority of scientists. We may ask how such an agreement is possible. The answer is easy to give: in cases of agreement either no or relatively few disconfirmations of the hypothesis under consideration are known, or if more disconfirmations exist, they are played down by the scientific community. Thus, in the latter case, the social force of the scientific community and the more or less pronounced conformist attitude of many scientists facilitate acceptance of the status quo with regard to scientific hypotheses, theories, and conceptual frameworks. It is also important to realize that often disconfirmations are ignored or suppressed. They are not mentioned in many textbooks and therefore many students may not realize that the hypotheses are actually questionable. At a later stage in their career, when the students or scientists may discover the suppressed disconfirmations, conditioning by the scientific community may already have reached a stage in which their minds have become so closed that certain disconfirmations no longer bother them seriously. Naturally, this need not be so, but a knowledge of the actual practice of science shows that these heretical-sounding statements are no exaggeration. I shall return to this topic in the following sections on the systems model of scientific methodology and on scientific progress. For concrete examples the reader is referred to Chap. 5 on plant morphology.

Before the systems model is presented, some more comments on internalism and externalism are necessary. The ongoing debate between internalists and externalists has polarized the issue considerably. Extreme views are therefore not seldom held. On one side, there are still authors who insists that psychological and social factors do not (or should not) influence the acceptance or rejection of scientific hypotheses. As has been pointed out already and as will be discussed in more detail in connection with the systems view of science, this extreme internalism is no longer tenable. On the other side, there are authors who underestimate the so-called internal factors of science and who therefore tend to an extreme externalistic view of science. Such extreme views are psychologism and sociologism of science.

According to psychologism only a psychological analysis of the scientific process will give us an insight into why one hypothesis is rejected and another one accepted and why science evolves in a certain way. This extreme view appears to disregard that in addition to psychological factors whose elucidation is the task of psychology, logical and empirical elements play an important role in scientific methodology.

Sociologism claims that only social analysis can provide insight into the scientific enterprise. Thus social factors are held completely responsible for the acceptance or rejection of hypotheses and for the progress of science. This extreme view has serious implications for the individual scientist as well as for society. If science, as sociologism claims, has no rational and objective guiding principles of its own, then its direction of development must be determined by society. More specifically, society has to decide whether a particular hypothesis is accepted or eliminated. Depending on the structure of society, this kind of decision would be made democratically or by a dictator or monarch. Thus science becomes politics. The question of which hypothesis or theory approximates reality more closely is no longer important. Since science has no objective and rational means of deciding between competing hypotheses or theories (for further reading on this problem see, e.g., Habermas 1970; Roszak 1973; Rose and Rose 1976a,b).

I presume that most if not all scientists protest against this extreme view of science. Scientists are simply too much aware of the importance of facts, logic and intuition. It is difficult to imagine that all facts, logic and intuition should be solely determined by society and in no way reflect nature. Our upbringing in a certain society colors our picture of nature, but nature also puts constraints on our confirmed hypotheses. Because of such constraints it is difficult to agree with extreme relativists such as Pearce (1973) who claim that any hypothesis is just as good as any other one because one can reach agreement between any hypothesis and facts simply by changing the auxiliary assumptions and our state of consciousness.

Lysenkoism is a well-known example of a theory or theoretical framework that was imposed on science from outside, i.e., a totalitarian state. Did it fail eventually simply because the political constellation changed – which would be the implication of sociologism – or because nature disconfirmed it? I tend to think that a considerable amount of empirical evidence points against Lysenkoism and therefore it is at least at the present time unlikely that Lysenkoism is correct (see, e.g., Lewontin and Levins 1976). However, we could not state that Lysenkoism has been falsified in a strict sense because we cannot exclude the possibility that new empirical evidence of the future might put Lysenkoism into a different perspective.

In conclusion, we may state that although psychologism and sociologism appear to be extreme views which appear to be unacceptable, they have the merit that they force us to reconsider the importance of psychological and social factors in scientific methodology. It seems that many scientists tend to naively underestimate how their scientific work is influenced by what they call outer-scientific factors. Very often we are not aware of our social conditioning and therefore we tend to believe that what we observe is "out there" in the world, whereas to a great extent it may be a projection of what is in us and what goes into us through our upbringing and education in a certain society.

## 1.3.4 A Systems Model of Scientific Methodology (Laszlo's Model)

The task of modern philosophy of science whose aim is the reconstruction of scientific methodology must be a comprehensive view of the scientific enterprise. It must take into consideration all factors that influence the acceptance or rejection of hypotheses and theories and must attempt to understand the mutual interaction of those factors. I think that systems models come closest to such a full representation of scientific methodology (see, e.g., Laszlo 1973; Radnitzky 1974; Bunge 1977).

In the following I shall present the "General systems model of the evolution of science" by Laszlo (1973) in a version that I have slightly modified and expanded by the addition of more feedback loops (Figs. 1.7).

The model of Fig. 1.7 may be looked upon as complementary to the scheme of Fig. 1.6. The latter shows only how hypotheses are confirmed or disconfirmed whereas the systems model addresses itself to the crucial ques-

32

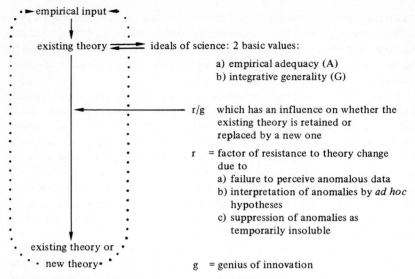

Fig. 1.7. Scheme illustrating the factors that influence theory change and tenacity to existing theories [modified after Laszlo (1973), p. 388]

tion of which factors determine retention or replacement of a hypothesis (or theory).

Three principal groups or kinds of factors are singled out:

(1) the data (= empirical input)
(2) the ideals of science, particularly in the form of the balance of A/G (empirical adequacy/integrative generality)
(3) the mutual relationship of resistance factor (r) and the genius of innovation (g).

The facts relevant to an existing theory (= empirical input) are of utmost importance. However, facts alone do not determine whether an existing theory is retained or rejected. Ideals of science, which can be looked upon as values of science, also have a considerable influence. Laszlo (1973) distinguishes two basic values: empirical adequacy (A) and integrative generality (G). Other values could be added (see, e.g., Mohr 1977, lectures 11–14). "Empirical adequacy is a measure of the number of facts accounted for by the science, and the precision, detail, and predictive power whereby it provides its account" (Laszlo 1973, p. 388). In short, empirical adequacy refers to the agreement of facts and theory. The greater this agreement, i.e., the more confirming facts and the less disconfirming facts we have, the greater is the empirical adequacy. "Integrative generality (G) is a measure of the internal consistency, elegance, and neatness of the explanatory framework. It is determinable in reference to the number of separate assumptions made in a theory concerning its subject matter. Generality increases proportionately to the range of application of the basic existential assumptions and hypotheses. The smaller the number of such hypotheses in relation to a larger number of facts explained, the higher the generality of the theory" (Laszlo 1973, p. 388). Integrative generality is culture dependent. Social as well as psychological factors play a role here. An example is the desire to have elegant theories. The notion of elegance itself may depend on social and psychological conditioning. Thus, what to one person is elegant, to another one may not be elegant or less so.

According to Laszlo a balance of empirical adequacy (A) and integrative generality (G) occurs when "the optimum precision is reached with regard to the largest number of facts by deduction from the smallest number of existential assumptions" (Laszlo 1973, p. 388). It is not easy to reach such a balance and I also doubt whether it can be determined objectively. The fact that empirical adequacy (A) and integrative generality (G) may be antagonistic to each other complicates the matter. A theory with a high degree of empirical adequacy may have a low degree of integrative generality (G) and vice versa. One investigator, one or a few powerful scientists, or the majority of the scientific community at a certain time may have a preference for A or G in cases of conflict. This shows how the acceptance of ideals of science (particularly the mutual relationship of A and G) may influence whether a theory is retained or replaced. If empirical adequacy is more highly valued then integrative generality, a few conflicting facts may lead more easily to the overthrow of a theory than in the opposite case in which preference is given to internal consistency, elegance, and neatness.

However, the process of theory retention and replacement is more complicated than that. It is greatly influenced by the mutual relationship of r and g. r is "the factor of resistance to theory-innovation due to textbook indoctrination with a reigning paradigm (resulting in a failure to perceive anomalous data; its interpretation by ad hoc auxiliary hypothesis; or its suppression as a temporarily insoluble problem)" (Laszlo 1973, p. 391). The importance of this resistance factor is often underestimated by scientists because, to a great extent, it works subconsciously as it is characteristic of indoctrination (for example see Barber 1961). Many scientists are not aware of all the facts that contradict a given theory. And those contradictory facts that are more generally known are often "explained away" by ad hoc assumptions or are not taken seriously. Thus a theory can survive and even flourish in spite of the existence of contradictory facts.

The scientific community does not easily give up a theory. In this sense, the scientific community has a strong conservative tendency. Feyerabend

(1975) called this clinging to an existing theory the "principle of tenacity." Evidently, tenacity can create a considerable obstacle to change and progress in science. However, it may also have a positive aspect because it forces the scientific community to explore more fully the validity of a theory before it is given up (see, e.g., Kuhn 1977). History of science provides examples of facts which at first appeared contradictory to the existing theory and then at a later stage could be explained by the same theory (see, e.g., Hempel 1966, p. 52; but see also p. 54). In this way the unnecessary proliferation of too many competing hypotheses and theories is avoided. Feyerabend (1975) contrasted the "principle of proliferation" to the "principle of tenacity." Obviously both principles play a role in the scientific enterprise. It seems, however, that in many cases tenacity is stronger, especially with regard to the most fundamental theories including ways of thinking.

As the number of contradictory facts increases, the likelihood that the theory will be given up also increases. However, normally even a rather unsatisfactory theory is not rejected unless someone has proposed an attractive alternative. Hence, g, the genius of innovation is of utmost importance with regard to theory replacement. One does not want to give up a theory for nothing at all. Likewise, many people who are dissatisfied with their husbands or wives do not reject them completely unless they have a new, more attractive lover. The feeling (or illusion) of security is as widespread in everyday life as in science.

Laszlo (1973, p. 391) points out that as a result of the resistance factor there are thresholds of tolerance. These thresholds are a function of both the resistance factor (r) and the availability of attractive alternatives (g). The two factors are inversely related: the higher g, the lower r, and vice versa.

Laszlo's model shows the multifactorial nature of theory retention and replacement, taking into account the empirical basis, logical argumentation (deductive reasoning) as well as psychological and social factors. The latter may introduce nonrational aspects. Hence, the scientific enterprise cannot be characterized as totally rational. A theory is neither retained nor rejected for solely rational reasons. As pointed out, scientists may cling to theories even in the presence of contradictory data. The reasons for such tenacity may reach deep into the subconscious.

Rationality and logic as well as the empirical basis of the data play an important role in science. However, irrationality or nonrationality cannot be completely discounted (e.g., Newton-Smith 1981; Feyerabend 1982). This poses a problem with regard to the demarcation of science toward nonscience such as philosophy, religion, or the arts. Many attempts have been made to provide criteria for a clear-cut demarcation. Laszlo (1973, p. 388) uses the balance A/G as the important criterion. A scientific theory exhibits such balance, whereas an unscientific theory does not.

However, since the ratio of A/G is a matter of degree, I think that the difference between a scientific and a nonscientific theory is also a matter of degree. Thus, a continuum ranging from strictly scientific to completely nonscientific postulates can be recognized. Consequently, the attempt of demarcating science from nonscience is illusory (see also, e.g., Bunge 1977; Hull 1983). This includes the methodology. Hence, a methodology absolutely unique to science does not exist because scientific methodology intergrades with nonscientific methodology. What I have called "scientific methodology" is typical for science, but it is not totally discontinuous with methodologies used in everyday life and activities which are normally classified as non-scientific (e.g., Feyerabend 1975, 1982).

Laszlo's model is comprehensive. Yet I think that additional complexities occur which I indicated by stippled arrows in Fig. 1.7. As will be pointed out especially in the chapter on facts, facts are theory dependent. Thus, a new theory may lead to a different perception of facts. Furthermore, a new theory may influence the ideals of science, and the ideals of science in turn may affect the perception of facts. This shows that the scientific enterprise is a highly complicated network of interrelations between facts, logic, psychological, and social factors. We may ask whether under such conditions progress is still possible. In other words: does the acceptance of a new theory necessarily mean that it is closer to the truth than its rejected predecessor, or does the theory change rather reflect a psychological/social preference?

## 1.4  Is Scientific Progress Possible?

Much controversy exists on the question of whether scientific progress is possible, and if so, to what extent and in what sense (see, e.g., Kuhn 1970, 1977; Lakatos and Musgrave 1970; Jevons 1973; Stegmüller 1976, 1977; Gutting 1980; Dilworth 1981). Laszlo (1973) believes that his "General systems model of scientific evolution" can support the idea of progress in science. He considers science as an open system analogous to the open system of an organism. As the organism (or rather the species) adapts itself increasingly to its environment, so science enhances its adaptation. The environment of science is nature. An increasing adaptation of science (i.e., scientific theories) to nature means that they approximate more and more the truth. "The 'ideal' or 'perfect' science would be fully adapted to nature, i.e., it could explain and predict all empirical inputs" (Laszlo 1973, p. 387).

Although Laszlo provides a theoretical foundation for the contention of progress in terms of General Systems Theory (Laszlo 1972a, 1974), the question is how increasing improvement of scientific theories can be assessed. A number of difficult problems confront us in this respect. One of them is the incommensurability of theories and paradigms (Kuhn 1970; Nersessian 1982). In order to discuss this problem it is useful to describe some of Kuhn's (1970) ideas.

Kuhn (1970) makes a fundamental distinction between "normal science" and "revolutionary science." "Normal science means research firmly based upon one or more past scientific achievements" (Kuhn 1970, p. 10). More specifically, these achevements must be of a twofold nature: (1) they must attract a scientific community away from competing modes of scientific activity, and (2) they must be sufficiently open-ended to leave all sorts of minor problems to be solved, adjustments to be made, re-definitions to be introduced, etc. Achievements that share these two characteristics are called paradigms. One of Kuhn's critics pointed out that Kuhn used the term "paradigm" in at least 22 different ways in the first edition of his book (see Kuhn 1970, Epilogue, p. 181). Therefore, Kuhn (1970, Epilogue) made the distinction between "exemplar" and "disciplinary matrix." An exemplar refers to problem solutions that we encounter in various scientific contexts (see Kuhn 1970, p. 187). A disciplinary matrix is, among other features, characterized by a system of beliefs, values, and symbolic generalizations (Kuhn 1970, Epilogue). A notion related to that of "disciplinary matrix" or "paradigm" is Rozov's "normative system" which is "the totality of models (examples, patterns) by means of which any human activity, including thought is exercised" (Meyen 1982).

Although Kuhn (1970) has made the distinction between "exemplar" and "disciplinary matrix", the term "paradigm" is still used for both of these concepts and a variety of related concepts (e.g., Simberloff 1980; Grene 1980). It is not clear which (if any) paradigms may be distinguished in biology. Table 1.1 presents an example of three paradigms. Most biologists still oscillate between all of these three paradigms, which makes their distinction difficult, if not impossible.

Revolutionary science is the replacement of one paradigm or disciplinary matrix by another (e.g., Hacking 1981; Cohen 1985). This involves the replacement of one major theory by another. Now the crucial question with regard to scientific progress is the following: Are the new theory or the new disciplinary matrix superior to the old one, i.e., does the revolution represent progress? It is often said that the new theory is more comprehensive than its predecessor; therefore it can explain and predict a wider range of phenomena. Does this mean that it comes closer to the truth, thus constituting progress? Kuhn has emphasized that theories of different disciplinary matrices are incommensurable, i.e., they are so different in their conceptual framework and their domain of applicability that a comparison is impossible. How then is it possible to decide which one is better? It has been said that this whole question is misdirected because of the lack of comparison. According to this

Table 1.1. Characteristic of three paradigms (after Maruyama 1974, modified by Johnson 1977). The three paradigms "correspond roughly to what we might loosely call "hierarchists", "individualists", and "mutualists" (Maruyama 1974) [Reported from Johnson (1977) by permission of the author]

| | Unidirectional causal paradigm | Random process paradigm | Mutual causal paradigm |
|---|---|---|---|
| Social organization: | hierarchical | individualistic | non-hierarchical inter-actionist |
| Ethics: | competitive | isolationist | symbiotic |
| Philosophy: | universalism | nominalism | network |
| Perception: | categorical | atomistic | contextual |
| Logic: | deductive, axiomatic | inductive, empirical | complementary |
| Science: | traditional "cause" and "effect" model | thermodynamics; Shannon's information theory | post-Shannon information theory |
| Research hypothesis and research strategy: | dissimilar results have been caused by dissimilar conditions. Trace to conditions producing them | there is probability distribution; find out probability distribution | dissimilar results may come from similar conditions due to mutually amplifying network. Network analysis instead of tracing of the difference back to initial conditions |
| Methodology: | classificational, taxonomic | statistical | relational, contextual analysis, network analysis |
| Information: | past and future inferrable form | information decays and gets lost; blueprint must contain more information than finished product | information can be generated. Nonredundant complexity can be generated without pre-established blueprint |
| Knowledge: | believe in one truth. If people are informed, they will agree | why bother to learn beyond one's interest | Polyocular: must learn different views and take them into consideration |
| Analysis: | pre-set categories used for all situations | limited categories for his own use | changeable categories depending on situation |
| Assessment: | "impact" analysis | what does it do to me? | look for feedback loops for self-cancellation or self-reinforcement |
| Decision process: | agency dictated | entrepreneur | participatory planning and evaluation |
| Esthetics: | unity by similarity and repetition | haphazard | harmony of diversity |

view the new disciplinary matrix, exemplar, and theory are just different, representing another perspective of nature. This conclusion shatters the belief in scientific progress at the level of disciplinary matrices. Two objections may be raised, however. First, as Kuhn himself (1970, Epilogue) has pointed out, theories of a new disciplinary matrix may allow more accurate predictions and the solution of a greater number of problems ("puzzles"). Thus, in this respect they may be superior to their predecessors in spite of their incommensurability. Second, it is questionable whether normal and revolutionary science can be easily distinguished in all disciplines. Toulmin (1974), for example, argues that Kuhn has exaggerated the discontinuities between disciplinary matrices. Greene (1971, 1981) provided an example by showing that the change from creationist to evolutionary thinking was a gradual process, not a sudden revolutionary shift. One may have to accept the idea then that at least some of the scientific shifts are continuous and thus do not correspond to Kuhn's model of revolutionary science. Probably a whole continuum from sudden (revolutionary) to gradual (evolutionary) shifts can be found in the history of science.

One might question whether in the case of a gradual shift the notion of incommensurability still applies. I think that in spite of the continuous change the difference of the early and late view may be so striking that they may indeed be incommensurable. Hence, even a gradual change of scientific thinking and theorizing would not necessarily remove the problem of incommensurability.

In discussing the possibility of scientific progress, another aspect has to be taken into consideration. When an evolutionary or revolutionary paradigm shift occurs, the new paradigm is not necessarily accepted by the whole scientific community. A split may occur in the community, part of it embracing the new paradigm and another part adhering to the old one. In this way more than one paradigm may coexist at a particular time. Each paradigm with its supporters represents a school of thought. This phenomenon of different schools is especially pronounced in psychology where today we find behaviorists, cognitive psychologists, Gestalt psychologists, psychoanalysts, etc. side by side (see, e.g., Brandt 1982). Since these schools are based on different paradigms (disciplinary matrices), meaningful communication between them is difficult (if not impossible). Thus the different schools continue their research more or less in isolation from each other. Some of them may have relatively few adherents, but that does not necessarily mean that they are further from the truth. We cannot tell whether they will attract more supporters in the future or whether they may become extinct with the death of their leaders. Even extinction is not necessarily an indication of the inferiority of the doctrine; it might reflect a lack of openness of the other school(s) and thus a failure to appreciate a highly evolved paradigm.

The coexistence of different schools of thought occurs in biology too, although probably to a lesser degree than in psychology. An example is the individualistic and the organismic views of plant communities (see, e.g., Yarronton 1967). According to these views a plant community is either an assemblage of individuals or a superorganism. The latter view has given rise to what is sometimes called the ecosystem paradigm according to which the ecosystem is seen as a "holistic unit of coevolution" (Patten 1975) that need not represent a superorganism (see Simberloff 1980). The ecosystem paradigm corresponds roughly to the mutual causal paradigm of Table 1.1.

From the point of view of scientific progress it is not satisfactory to have coexisting schools of thought. In order to make progress not only within one school of thought but beyond schools it would be necessary (1) to show that one of the schools is right and the others are wrong, or (2) to develop a new disciplinary matrix that unites those of all the coexisting schools. As discussed above, the first solution is unattainable. Nonetheless, theories may become extinct because their supporters die; but they may be revived in a modified form at a later time when the intellectual climate may be more favorable. The second solution, which aims at unification, is difficult to achieve. At present one may have doubts in many areas whether it will ever be possible. Yet one cannot be dogmatic in this respect and the unification of major theories and disciplinary matrices will remain a major challenge in science. Even in physics unification is not easy and it is debated whether all of the so-called unified theories have successfully assimilated other theories as special cases.

Since both elimination and unification of theories pose fundamental problems, the notion of progress in terms of improvement of theories or paradigms is also problematic. This conclusion need not lead to pessimism and resignation. Regardless of whether progress beyond paradigms is possible or not, we can learn much from scientific research. The development of a new paradigm can be a major enrichment of science and society. We have to accept, however, that such a development does not necessarily supersede completely existing paradigms. It presents a new perspective which, if not more comprehensive than previous ones, is complementary to them. Adding a complementary paradigm may also constitute progress, though progress in a different sense from that of improvement of a paradigm.

I think that the notion of complementarity is highly significant for science as well as society. It appears to be more realistic in many cases and it provides a basis for tolerance among scientists as well as nonscientists. A more general awareness of the complementarity of different schools of thought and of the problem of scientific progress may constitute progress in yet another sense (see also Chaps. 8 and 10).

### 1.5 The Semantic View of Theories

The way in which we view theories (and other theoretical constructs, such as laws) is also relevant to the question of progress (see Stegmüller 1976). Throughout this chapter I have referred to "theory" in a traditional sense, i.e., as "a system of logically related propositions" (Bunge 1980, p. 225); these propositions have empirical content, i.e., they tell us something about the world or some aspect of it. In contrast to this received view of theories, a semantic view has been developed by Suppes, Sneed, Stegmüller, and others (Beatty 1980, p. 400). According to this semantic view, a theory is only a definition; more specifically, it is "a statement that defines a type of natural system" (Giere 1979, p. 69). This means a theory no longer tells us anything about the world. In order to make a statement about the world we need a "theoretical hypothesis" which asserts that some specified system of the world is an instance of the type of system defined by the theory (Giere 1979, p. 70, 160).

If we now look at Mendel's theory of inheritance in terms of the semantic view, we have to recognize that it does not tell us anything about sweet peas or any other aspect of the world. It simply *defines* a system. Giere (1979) who developed a simple version of the semantic view of theories formulated Mendel's theory in the following way: "A system of inheritance of a single characteristic through sexual reproduction is a Mendelian system if and only if it satisfies the laws of segregation and dominance" (Giere 1979, p. 77). The two laws could have been written out in the definition of the theory. Only for the sake of clarity did Giere (1979, p. 77) state them separately. The important point is that the laws are also definitions without empirical content. Therefore, to make empirical claims we need theoretical hypotheses that pronounce actual genetic systems as instances of Mendel's theory. An example of such a theoretical hypothesis is the following: "The system of inheritance of height in sweet peas is a Mendelian system" (Giere 1979, p. 77). This example is a reformulation of one of Mendel's postulates in terms of the semantic view.

One major advantage of the semantic view of theories is that the endless discussion on the correctness, falsity, or the degree of adequacy and confirmation of any specific theory becomes irrelevant since the theory no longer makes claim about the world. The central question now is to what extent a theory is applicable, i.e., how many empirical situations are instances of the theory. When theories with an increased domain of applicability are developed, we may speak of progress. However, as long as there are situations that remain outside the domain of a certain comprehensive theory, the proposal of complementary theories that can absorb these situations also constitutes progress. Thus, the semantic view of theories is compatible with the idea of complementarity; and progress is not necessarily seen as a linear approximation to the truth (see Stegmüller 1976). The notion of truth is in fact altogether inappropriate with regard to theories and laws. Realizing this may have a rather liberating effect (see, e.g., Feyerabend 1975, p. 21). Since theories and laws do not make claims about the world, the question of whether there are laws in nature becomes also pointless from this point of view [see also Chaps. 2 and 8; for more examples on the semantic interpretation of theories and laws see Chaps. 5 and 8].

## 1.6 Conclusions

Modern philosophy of science has gone far beyond the naive belief that science reveals the truth. Even if it could, we would have no means of proving it. Certainty seems unattainable. All scientific statements remain open to doubt [see, e.g., Prigogine and Stengers (1979) who emphasize the openness of science]. We cannot reach the absolute at least as far as science is concerned; we have to contend ourselves with the relative. Nature and logic are not the only determinants of the evolution of science. Ideals and values, psychological and social factors (also called external factors) play an important role. Science is culture dependent. Subjectivity cannot be completely eliminated and irrationality or nonrationality may influence the scientific enterprise. Whether scientific progress occurs is questionable at least to the extent that paradigms may be incommensurable. However, the development of complementary paradigms may be seen as progress too. Inasmuch as the unity of science is not within reach, we have to content ourselves with a range of complementary perspectives (perspectivism). Thus science is pluralistic with regard to paradigms.

These conclusions need not lead to pessimism and resignation. Inasmuch as they are more appropriate than absolutistic, rationalistic, and objectivist dogmas, they provide a sounder basis, may cure our hubris, may lead to more humility, tolerance, and understanding and thus may diminish repression and destruction.

## 1.7 Summary

*Introduction.* Fundamental questions on scientific methodology are posed.
*The aim of science* is to gain scientific knowledge of the world, in the case of biology, scientific knowledge of the living world. Scientific knowledge is expressed in terms of singular and general propositions. Singular propositions are also called facts or data. General propositions comprise various kinds of

hypotheses, models, laws, rules, and theories. The ultimate aim of science is the acquisition of the most comprehensive and best confirmed theories (compare, however, below *The semantic view of theories*).

## Scientific methodology
*General considerations.* The importance of a knowledge of the methodology employed is emphasized. Laboratory methodology is used as an example to show that the perception of a cross-section of a leaf under the microscope is determined not only by the structure of the leaf but also by the preparation methods, the methods of microscopy, and the mode (method) of perception which depends on the organization especially of our sensory apparatus and nervous system. Hence, knowing one particular object requires an understanding of the methodology employed which includes the observer. In contrast to the methodology required for the perception of singularities, the term 'scientific methodology' usually refers to the methodology by means of which general propositions of science are obtained.

*Induction and the hypothetico-deductive method.* General propositions may be obtained through inductive generalization. The famous white swan example is used to show that this methodology has no logical force and cannot lead to certainty. The hypothetico-deductive method requires intuition, logic (deductive reasoning) and empirical input. General propositions, according to this methodology, are intuitive conjectures which are tested through a comparison of deduced and observed states of affairs.

*Validation of hypotheses.* The question is how intuitive conjectures can be validated. Is it possible to attain certainty either of the truth or falsity of an hypothesis or theory? Three postulates (philosophies) present answers to this question:
    *The first postulate of validation* implies that hypotheses may be proved. Hence progress in science is undeniably straightforward and cumulative. If we do not reach the total truth, we shall at least come closer and closer to it. This belief may lead to hubris, intolerance, repression and destruction. Those who believe that they possess the truth not seldom tend to impose their truth on others, if necessary by force. It is shown that the assumption of the first postulate of validation is not tenable: hypotheses cannot be proved.
    *The second postulate of validation* implies that hypotheses cannot be proved, but may be disproved (falsified). This postulate accounts for scientific revolutions as a replacement of one theory by another one. It reminds us that we may be wrong. However, the belief that we can show the falsehood of a hypothesis with absolute certainty cannot be defended. Two major reasons are given why the second postulate of validation is untenable. Three

versions of Popperian philosophy are presented the latter of which approaches the third postulate of validation.

*The third postulate of validation* implies that an hypothesis can neither be proved nor disproved; it may only be confirmed or disconfirmed. The burning question that arises then is how we can accept or reject any hypothesis? According to extreme views this is only possible through outer-scientific (external) factors. Psychologism claims that there is only a psychology of discovery and scientific progress. Hence, whether a hypothesis is accepted or rejected, is determined by psychological factors. According to sociologism, only social factors can explain the acceptance or rejection of hypotheses.

*A systems model of scientific methodology.* Systems models of scientific methodology present a broader view. They take into consideration internal (inner-scientific) as well as external (psychological and social) factors. Laszlo's (1973) "general systems model of the evolution of science" is presented in a slightly modified and expanded version (Fig. 1.7). According to this model the following three kinds of factors determine the acceptance or rejection of a theory: (1) empirical input (facts, data), (2) ideals (values) of science, and (3) the relation of a resistance factor to innovation. Resistance is due to indoctrination which may lead to the suppression of contradictory data, its explanation by ad hoc hypotheses, or its elimination as temporarily insoluble. The three kinds of factors interact with each other and with theories. Consequently acceptance and rejection of theories is influenced by empirical inputs, logic, psychological and social factors, i.e., objective and subjective factors, rational and nonrational aspects. Science is not totally rational. It is a complex enterprise which is not sharply delimited from nonscience.

*Is scientific progress possible?* Laszlo (1973) believes that progress is conceivable according to his systems model. Kuhn (1970) emphasizes the incommensurability of paradigms which means that they are not comparable. This complicates the assessment of progress when different paradigms are involved. It is questionable whether paradigms are as discontinuous as Kuhn assumes. However, the coexistence of different paradigms which creates schools of thought also seems to indicate that progress may occur only within a paradigm or disciplinary matrix. One may doubt whether this conclusion is inevitable, but even if it were, this would not be reason for pessimism and resignation. The development of a new paradigm or disciplinary matrix can be seen as an enrichment to science inasmuch as it presents a complementary perspective. The opening up of new perspectives is progress in a different sense from that in terms of improvement and unification of paradigms and disciplinary matrices. Finally, the awareness of complementarity and limits of scientific progress which may lead to more tolerance represents progress in yet another sense.

*The semantic view of theories.* According to this view, theories and laws are only definitions, i.e., they are not about the world. Therefore, the question whether a theory or law is correct, false, confirmed or disconfirmed, is no longer meaningful and is therefore replaced by the question whether a theory or law applies to a particular situation. If more than one theory or law applies, these are considered complementary. It is obvious that the semantic view of theories is of fundamental importance to science and philosophy.

# 2 Laws, Explanation, Prediction, and Understanding

"Too often we forget that science evolved out of the rationalism of the Middle Ages, which was characterized by a
faith based on reason, and has become, instead, a reason
that is based on faith — faith in the ultimate orderliness of
Nature" (Davenport 1979, p. 2)
"Whenever we proceed from the known into the unknown
we may hope to understand, but we may have to learn at
the same time a new meaning of the word 'understanding'"
(Heisenberg 1962, p. 201)

## 2.1 Laws

A law, i.e., a scientific law of nature, may be defined as a statement describ-
ing a regularity of events or characteristics [see, e.g., Walters (1967) who also
discusses other views]. Thus, a law refers to a pattern which, according to
our experience, is invariant; invariant may mean universally invariant. A number of more detailed definitions of the concept of law have been proposed
(see, e.g., Ruse 1970), and different types of laws have been distinguished,
such as, for example, deterministic and probabilistic laws. A deterministic
law [which Hempel (1966) calls a 'universal law'] is a statement which says
that "whenever and wherever conditions of a specified kind F occur, then so
will, always and without exception, certain conditions of another kind, G"
(Hempel 1966, p. 54). In short: when F, then G. Although this definition of
a deterministic (or universal) law may appear satisfactory at first sight, it
raises problems for several reasons (see, e.g., Walters 1967). One of these
problems is that it is also true for accidental (or descriptive) generalizations
which are not laws. An example of an accidental generalization is the statement: all apples in this basket are red. Evidently this statement is not a law;
however, it satisfies the above definition: when F, then G – F being the condition of being an apple in the basket, and G the condition of being red.
Thus the characterization: when F, then G, is a necessary but not a sufficient
condition for the definition of deterministic laws. The question then is: how
can laws be defined in such a way that accidental generalizations are excluded?
As pointed out, for example, by Hempel (1966, p. 56) a law can support
counterfactual conditionals, whereas an accidental generalization cannot.
Counterfactual conditionals are statements of the form: "If A were (had
been) the case, then B would be (would have been) the case" (Hempel 1966,
p. 56). Thus, if any apple had been in the basket, it would not necessarily
have been red. Hence, the statement on apples in the basket being red is not

a law. However, the statement that mitosis divides the genetic material into equal parts is a law because it supports the counterfactual conditional that if mitosis were to occur, the genetic material would be divided into equal parts. [Concerning subjunctive conditionals which are also supported by laws see Hempel (1966), p. 56].

The criterion of counterfactual (and subjunctive) conditionals is not accepted by all philosophers of science as sufficient for the distinction of laws. Tondl (1973, p. 181) remarked that "there is no way of demonstrating any clear-cut or unequivocally definite boundary between expressions which we acknowledge as having the character of scientific laws, and expressions which we consider as empirical generalizations, etc." Nonetheless, the search for criteria of demarcation continued. Hull (1974, p. 71) thinks that the only criterion "that shows any promise of being adequate is the actual or eventual integration of natural laws into theories, while descriptive generalizations remain isolated statements." This approach has at least two consequences which may lead to difficulties: (1) it makes laws theory-dependent, and (2) "at any one time in the development of science it is not easy to decide which apparently true universal generalizations are merely descriptive or accidental and which will become incorporated into currently emerging theories" (Hull 1974, p. 71). Hull presented the following example of a generalization (regularity) which has not yet been incorporated into scientific theory: All the proteins that make up terrestrial organisms are of the same kind – the levo form (A few dextro amino acids are associated with certain bacteria)" (Hull 1974, p. 72). He noted that "although there are chemical reasons for all the proteins in a single organism to be of the same form, there is nothing about current chemical or biological theory that would lead one to expect this form to be levo rather than dextro. From the point of view of current scientific theory this fact is purely accidental" (Hull 1974, p. 72). Hull considers the concept "accidental" to be relative. This means that what may be accidental or random with regard to one body of laws and/or theories, may turn out to be determined in terms of a different set of laws and/or theories.

Mohr (1977, p. 62) lists a number of deterministic (universal) laws in biology. He refers to them as principles which, to him, is just another name for "universal law." I mention here only the first of his examples: $\Delta G \neq 0$, whereby G is the symbol for Gibbs free energy. According to another formulation of this law, "every living system requires the continuous supply of free energy to compensate for the continuous production of entropy" (Mohr 1977). This formulation says: "there is no living system that is in thermodynamic equilibrium" (Mohr 1977).

In contrast to a deterministic law, a probabilistic law is a statement that can be briefly formulated as follows: when F, then probably G. It can also support counterfactual and subjunctive conditionals. Many biological laws, such as for example Mendel's laws, are probabilistic (see also, e.g., O'Brien 1982; Knox and Considine 1982).

Both deterministic and probabilistic laws may be either process laws or coexistence laws. A process law implies that G follows F in time, whereas a coexistence law requires that F and G be contemporaneous. Mendel's laws, for example, are probabilistic process laws, whereas laws of comparative morphology are coexistence laws [see Chap. 5 and Mohr (1982)]. [For other examples of coexistence laws (or cross-section laws) and their critical evaluation in phylogenetic investigations see Hull (1974), pp. 77-80].

There are laws that may be comparable to axioms of a purely deductive system such as, for example, the law that "all physical laws (= laws of physics) may be applied to living systems (or, are valid in biology)" (Mohr 1977, p. 63). The validity of this kind of law is dependent on an all-inclusive philosophy or meta-theory of living systems (see Bunge 1977). Other laws that are propositions deduced from more general propositions depend on the proposition from which they are deduced in addition to the more inclusive philosophy mentioned above. Empirical laws according to Mohr are "general propositions that are results of limited inductive inference rather than of deductive reasoning" (Mohr 1977, p. 56). These empirical laws depend on the whole network of auxiliary hypotheses and the higher level theoretical framework of the particular discipline. They integrade with the more fundamental and general laws [for a discussion of causal laws, which are process laws, see, e.g., Hull (1974), pp. 72-87 and the chapter on causality in this book].

Even the most general and the most highly confirmed laws are subject to doubt. As is the case for any scientific proposition, we cannot attain certainty for laws. Hence we must remain open and be prepared that new empirical evidence may require an alteration of our laws. Although many scientists take this critical attitude toward our present-day laws, they often insist that there is an inherent order and lawfulness in nature. If our present laws are not perfect, they say, that does not indicate that there are no perfect laws in nature. It simply means that we have not yet found the laws of nature. But how can we be sure that there are laws in nature? The fact that we have been successful in formulating laws that, to varying extents, appear to correspond with nature can be interpreted in two ways: (1) It can mean that we have discovered in those laws the inherent order and lawfulness of nature; exceptions to those laws contradict, of course, this view. (2) It could mean that the laws are our inventions which correspond to nature only partially; exceptions do not contradict this view. An analogy may illustrate what is meant by a partial correspondence. It has been said that there are no pure circles in nature. Pure circles are only in our minds. Yet, although there are no (pure) circles in nature, projecting the notion of circle into nature does not lead to totally fictitious descriptions and predictions, but allows us to represent nature rather satisfactorily in many instances. If we arbitrarily limit the application of our notion of circle to structures that for all practical purposes approach a circle, we might be led to the conclusion that circles actually exist in nature. If, however, we extend our notion of circle to less circular structures that intergrade with the more circular ones, we can see more easily that we are projecting an orderliness into nature which is not there. Hence, instead of being led to the notion of order and lawfulness inherent in nature, one might be inclined to see chaos in nature. However, since a certain tendency of regularity may be noted in many instances, a totally chaotic view of nature seems to be at variance with the findings of science. It should be noted too that in some cases lawfulness has been detected in apparently chaotic phenomena (Wimsatt 1980; May 1981, Chap. 2, p. 14; Schaffer and Kot 1985). Such achievements may foster the belief in the lawfulness of nature. Nonetheless, one has to keep in mind that to date most if not all laws have exceptions and thus do not qualify as laws in the strict (universal) sense of the term [see, e.g., Beatty 1980; Stent's (1978, p. 55) discussion of second stage indeterminism is also relevant to this issue]. I therefore suggest that we look upon order and chaos as complementary views of nature. Each one may represent an aspect or perspective of nature, but neither alone represents nature as she is. Watching the waves of the sea can nurture within us the notion of regularity and orderliness. One can hardly deny that there is an aspect of lawfulness. Yet one also has to admit that no wave is like another and that at times it is doubtful whether the concept of wave still applies. Thus, nature is neither order nor chaos, she is the Unnamable. I cannot do better but to quote Laotse's opening paragraph of the Tao te Ching (Lin Yutang's edition):

> The Tao that can be told of
> Is not the Absolute Tao;
> The Names that can be given
> Are not the Absolute Names
> The Nameless is the origin of Heaven and Earth
> The Named is the Mother of All Things

Now we may look at an example of a universal biological law in the light of the preceding comments. The example is the law (or principle) of compartmentation which states that "every living system is compartmented" (Mohr 1977, p. 62). Organisms of the greatest diversity are composed of cells (whose original meaning is compartment or chamber). Cells are compartmentalized into various organelles such as nuclei, mitochondria, etc. Yet in spite of this ubiquitous compartmentation, the law referring to it represents at best one aspect. We can perceive living systems also as the opposite of a system of compartments, namely, a continuum. For example, cells are usually continuous with each other through cytoplasmic bridges (plasmodesmata in plant cells and gap junctions in many animal cells) (DeRobertis and DeRobertis 1980; Hagemann 1982; Gunning and Overall 1983). Thus, a traveler of molecular size can move through a whole multicellular organism without crossing any boundary. To such a traveler, a living system is continuous. Analogously, a mouse running along highways in Europe would not come across any barrier, but would move around in a completely continuous network of roads. Even the "iron curtain" or the wall surrounding West Berlin would be nonexistent to that mouse. Thus the subdivision of Europe into separate nations is just as much a one-sided point of view as the compartmentation of an organism into cells.

In addition to the continuum, discontinuities (i.e., compartmentation) may still exist to our imaginary cytoplasmic travelers because membranes such as those of the endoplasmatic reticulum may still be perceived as such provided the traveler has a relatively large macromolecular size. If, however, the traveler assumes a smaller size, (s)he may move directly through the walls of compartments without necessarily perceiving them as walls that separate one entity from another. Discontinuity and continuity are thus dependent on the size of the observer, or, for a given observer, on the distance from the observed. For example, from an airplane we may see a perfect discontinuity between a lake and surrounding forest. If, however, we walk from the forest into the lake or vice versa we may observe a very gradual transition from one to the other. In the same way, an atomic or molecular traveler may not encounter any barrier or wall while moving from one cell to another through the so-called wall or membrane. Thus, continuities and discontinuities are relative. In any particular case one may state that the system is compartmentalized and at the same time that it is not compartmentalized. These two statements, which appear contradictory, are two different perspectives that are complementary to each other. Quarreling about which one is true and which is false, does not advance our knowledge, but is a destructive impediment to understanding. However, in a particular context one may be more appropriate than the other.

In addition to the one-sidedness and relativity of the law of compartmentation, I want to draw attention to a third limitation. As pores in the walls of adjacent compartments become increasingly wider, we reach eventually a situation in which the "wall" is reduced to a net. Since a net is no longer perceived as a wall with pores (especially if the spaces within the net are very large), the original compartments disappear. Thus, we cannot apply the term 'compartment' any more. In intermediate situations it is rather arbitrary whether we speak of 'compartments with very large pores,' or a unit with interior frameworks or nets. Therefore, in these cases it is meaningless to ask whether the law of compartmentation applies.

In Western culture, which has led us to considerable scientific knowledge and technological control, great emphasis is placed on law and order. We may ask why this is so. I think there are historical reasons. Western culture has its roots in Judeo-Christian religion in which the law was God-given and all-important. Later on, God was eliminated by many philosophers and scientists, but the law remained. It was forgotten by many that the faith in lawfulness had its roots in the conception of a certain God. In other religions and especially in mysticism there may be much less or no emphasis on laws. If God is referred to at all, he is rarely envisaged as a provider of law. Rajneesh (1975c) wrote: "God is absolutely wild." Neither he nor his creation (if we want to make this unnecessary distinction) can be put into a mould of law and order. Therefore, profound wisdom cannot be an orderly expression either. It is rather "crazy wisdom" (Trungpa 1973, 1976). One might object here that I am confusing laws and order of nature with those of society. I think, however, that there is a closer connection between the two than is commonly realized. Science does not operate in a vacuum, but in a cultural context. Thus the widespread notion of law and order has become incorporated into its basic presuppositions.

Since neither the truth of scientific laws nor the truth of lawfulness in nature can be proven, the expectations of the scientist who hopes to discover the lawfulness in nature are based on belief or faith. One may argue that this belief has been confirmed, yet, in very general terms, it is still belief. It seems that this conviction (or this belief) in lawfulness has been a tremendous driving force in the evolution of science. And it also seems to be correct that for the majority of scientists this conviction is necessary in order to devote themselves to science. However, it is possible to do scientific research without this conviction. Especially in the twentieth century, science itself has led us to a recognition of its limitations and as a consequence at least some of us have become more modest (see, e.g., Stent 1978; McIntosh 1980/82). Thus it is less pretentious and less speculative to assume that instead of discovering the laws of nature we are inventing laws that correspond with nature to a certain extent; or in other words: that these laws represent only the lawful aspect of nature, not its chaotic or capricious aspect.

Riedl (1980) and other evolutionary epistemologists would contend that our urge for lawfulness is not only the result of cultural conditioning, but has deep biological roots in evolution. This contention is rather speculative and I think it is highly questionable. If it were correct, one should expect it to be universal. That is not the case. In both Western and Eastern cultures certain individuals have been free (or have freed themselves) of the urge for lawfulness. In fact, whole religous traditions such as Taoism have rejected the contention that there is lawfulness in nature. According to Taoism the supreme man is he who sees chaos in nature (Izutsu 1967). Chaos here means "a state of formless fluidity: nothing stable, nothing definite" [Chuang Tzu, quoted by Izutsu (1967), p. 402]. Today many people, especially within the counterculture, are rejecting or questioning their belief in law and order. This need not lead to anarchy. On the contrary, if it is grounded in a profound understanding and experience of life, it may create harmony.

To some extent, our modern Western society has become polarized between adherents of the counterculture and the traditional culture. Extreme, dogmatic, and destructive statements have been made on both sides. I think that both sides and society at large would greatly benefit from a recognition of the limitations of each viewpoint. Thus a reconciliation would be possible accepting that chaos and order are just different aspects of the Unnamable.

It should be added that the preceding conclusions depend on a certain conception of lawfulness and chaos. Lawfulness is understood in the sense of invariant regularities. That is the notion that, for example, is advocated by Hempel (1966) and many other philosophers of science. Chaos is understood as the negation of the above kind of lawfulness. Instead, it could be viewed as a state that lacks the simple lawfulness of "when F, then G", but nonetheless does not imply that everything is possible. In other words: chaos according to this other view may refer to an interdependence of events, yet not in a form that would correspond to any traditional notion of order and regularity (see also Chap. 8).

If the notion of lawfulness is defined in a broader way that does not require the "when F, then G" stipulation and if chaos implies interdependence, it is conceivable that these different definitions of lawfulness and chaos actually approach each other and that certain connotations of them may perhaps even coincide. In that case, the quarrel between those who advocate lawfulness and those who talk of chaos is ill-founded because it is based on a misunderstanding of the meaning of terms. Thus, proponents of different terms may not disagree as much as they think and as the terms seem to indicate.

## 2.2 Explanation

Explanations are important to most of us. As children we already notice phenomena for which we want explanations. We ask: Why is this so? Why did he die? Why do we suffer? And we expect explanations. The question that arises in the context of a scientific approach is the following: how does science explain? Or: what is a scientific explanation?

According to Hempel (1966, p. 48) scientific explanations must meet two requirements:

(1) the requirement of explanatory relevance, i.e., the propositions used for the explanation must be relevant [for an amusing example of explanatory irrelevance see Hempel (1966), p. 48]

(2) the requirement of testability, i.e., "the statements constituting a scientific explanation must be capable of empirical test" (Hempel 1966, p. 49).

As these requirements are met, an explanation is possible if the phenomenon to be explained (i.e., the explanandum) can be derived from an explanans (i.e., that which explains) that consists of a law (or laws) termed covering law(s) and the pertinent particular circumstances (initial conditions and/or boundary conditions). This model of explanation is referred to as the covering law model of explanation (Hempel 1965, 1966). Two types of explanation are distinguished depending on the kind of laws used for the explanation: (1) deductive-nomological explanation which is based on deterministic (= universal) laws ('nomological' is etymologically derived from 'nomos,' for law), and (2) probabilistic explanation which utilizes probabilistic laws.

1. *Deductive-nomological explanation.* — This kind of explanation is a deductive subsumption under deterministic (universal) covering law(s) and fact(s). What has to be explained (i.e., the explanandum) follows logically as a special case of a covering law (or laws) and relevant particular circumstances. Thus, "the explanation fits the phenomenon to be explained into a pattern of uniformities and shows that its occurrence was to be expected, given the specified laws and the pertinent particular circumstances" (Hempel 1966, p. 50).

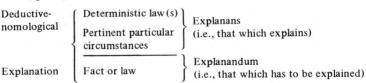

The explanandum may be a fact or a law. If it is a law, then the covering law(s) has (have) to be more general than the law to be explained. Mohr (1982) gave the following example of a deductive-nomological explanation:

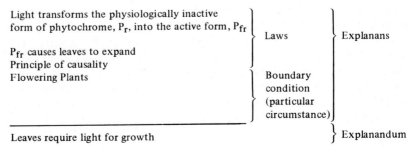

This example shows that in addition to the two laws the general principle of causality is implied (for a criticism of this principle see Chap. 6). As a result of the boundary condition, the validity of the two more specific laws is restricted to flowering plants.

Since in the covering-law model of explanation law(s) and particular circumstance(s) are necessary, a crucial insight of an explanation may be due to the discovery (invention) of a new law (or laws) and/or circumstance(s). Auxiliary assumptions may also be necessary. Furthermore, all sorts of other factors and background knowledge which are usually not stated explicitly may be relevant to the explanation (see, e.g., Wimsatt 1976, 1980; Hull 1981a).

2. *Probabilistic explanation.* In the case of a probabilistic explanation the covering law(s) are probabilistic. Hence the explanandum cannot be deduced with logical force, but can only be inferred with high probability according to the following scheme:

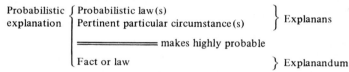

Hempel (1966, p. 59) gives the following example of a probabilistic explanation:

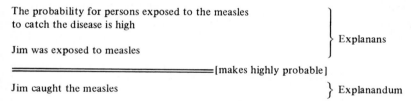

Since many biological laws are probabilistic laws, many explanations in biology are of a probabilistic nature, i.e., they are not arrived at with deductive certainty, even if the covering law(s) and fact(s) would be certain (see, e.g., Knox and Considine 1982). However, since neither laws nor particular circumstances can be certain, all explanations, including deductive-nomological explanations in which deductive certainty exists, are of a hypothetical nature, i.e., they might be wrong.

It has been debated whether the covering law model of explanation is the only way of explanation. If this were so, then areas of research in which laws cannot be formulated could not provide explanations. Areas of such research might be historical disciplines such as phylogenetics, although it is debated whether the difficulty in obtaining laws is inherent in the disciplines or simply due to lack of information at the present time. Regardless of the reason(s) for the paucity or lack of laws in such disciplines, we might ask whether it is possible to have a model of explanation that does not depend on covering laws. Goudge (1961) and a number of other authors (see, e.g., Weingartner 1967) answered this question in the affirmative. Alternative models of explanation proposed by Goudge are "integrative explanation" and "narrative explanation" (see Goudge 1961; Hull 1974, 1981). In the latter kind of explanation, "an event like the evolution of mammals or extinction of dinosaurs is explained by specifying the temporal sequence of events which lead up to it" (Hull 1974, p. 97). "Laws or law-like statements are explicitly said not to play a role. The event to be explained is fully explained by virtue of the fact that it falls into place" (Weingartner 1967, p. 9). Ruse (1973) objected to this claim by pointing out that theories, such as evolutionary or genetic theories, play a central role in such explanations. Hull (1981) agreed, but added that their "role is not the derivation of the event to be explained." According to Hull, theories and laws are mainly used to determine the historical entities (such as dinosaurs) or particular individuals for which a narrative explanation is given.

Woodger (1967) characterizes explanation in a very general way as analysis or relating. "The process of explanation will consist, therefore, either in exhibiting the relation of what is to be explained to something else, or in diminishing its complexity by analyzing it" (Woodger 1967, p. 273). Woodger doubts whether it is possible to achieve a purely analytical explanation and suggests that in the actual procedure of science both analysis and relating are always operative.

Jeuken (1968) also supports a broad view of the notion of explanation which to him is a synonym of "clarification" (van Laar and Verhoog 1971, p. 289). In his opinion explanation is "everything that contributes to our insight into the phenomenon under study." As a consequence of this broad view he distinguishes eight different kinds of explanation (see van Laar and Verhoog 1971, p. 288). Tondl (1973) and Achinstein (1983) also describe a variety of different models of explanation.

Explanation is the product of an explaining act, i.e., verbal or written communication. According to Achinstein (1983), the explaining act has been neglected in most models of explanation. Thus, Hempel (1965, 1966), like many other explanation theorists (see Achinstein 1983), stressed mainly the product of explanation and its logic. Achinstein tries to show that an explanation cannot be understood independently of the explaining act (see, e.g., Achinstein 1983, pp. 81-83). This is so because "what explains something for someone might not explain it for someone else, because of differences in beliefs, puzzlements, or intelligence" (Achinstein 1983, p. 9). Thus, the persons involved in the process of explaining become important and explanation can be seen as relative to these persons and their context including their cultural milieu. Psychological and social factors are again involved (see Sects. 1.3.3/4).

## 2.3 Prediction

Prediction in the strict sense is the forecasting of future events, whereas prediction in the wide sense is the forecasting of any unknown event, i.e., it comprises postdiction or retrodiction which is the inference of events of the past, such as the existence of certain fossils before they were actually discovered. To some extent the pattern of prediction in both senses is comparable to that of the covering law model of explanation. In both cases the premises consist of law(s) and particular circumstance(s). However, the deduction functions either as an explanation or a prediction depending on whether the deduced fact is known (as in the explanation) or unknown (as in the prediction).

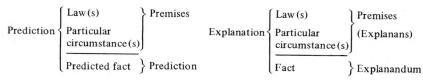

Although the symmetry between the two models appears striking, more and more asymmetries have been discussed in the literature (see, e.g., Hull 1974, p. 92). Whereas some of these asymmetries may apply generally, others may occur only in certain kinds of predictions. An example of the latter kind was mentioned by Mohr (1977, p. 69): in the case of X-ray induced point mutations precise prediction is not possible because such mutations are probably always quantum events which belong to the realm of microphysics where the uncertainty principle applies. Explanation of such a mutation (once it has occurred) is, however, possible in a precise way.

Some authors place enormous emphasis on prediction and the predictive power of hypotheses (and theories) and go so far as to suggest that general scientific propositions are defined in terms of predictive power, which means that only a proposition with predictive power is scientific. It is furthermore often assumed that the greater the predictive power of a hypothesis, the higher its degree of confirmation and the closer it is to the truth. Although it may be tempting to embrace prediction and predictive power as *the* criterion for science, some warnings may be appropriate. On the one hand, it is known that false hypotheses (i.e., hypotheses which we think are false) may have considerable predictive power. A well-known example is Ptolemaic astronomy which allowed and still allows accurate predictions of the movements of stars. On the other hand, correct hypotheses coupled with wrong auxiliary assumptions might have no or low predictive power. For these reasons it seems questionable to defend predictive power as the sole criterion for the validity of a scientific hypothesis, law, or theory.

Furthermore, our scientific understanding of life has revealed limits of predictability which makes prediction difficult or impossible in many instances (e.g., Hertel 1980; Rigler 1982). Hertel (1980) wrote: "We know of death by accident, of chance distribution in meiosis, of genetic drift, and many similar events which are practically unpredictable because of complex mechanism and/or small numbers. Furthermore, analyses in order to predict may interfere with, damage, or kill the organism, and beyond all this the "real chance" of quantum physics is very relevant for biology." Hertel concludes: "The death of Laplace's demon (i.e., the complete predictability of all events) is a happy result of science, and one should not try to correct nature by artificially recreating a predictable nature."

Kochanski (1973) analyzed the conditions and limitations of prediction (forecasting of future events) in biology. He distinguished two sources of difficulties: (1) general laws and theories, (2) initial conditions.

To 1. As pointed out already, general propositions such as laws or theories are a prerequisite for successful prediction. For example, if a process law specifies "when F, then G", we can predict that whenever we observe F, it will be followed by G. Difficulties in prediction may arise because of exceptions to the regular connection of F and G. The latter are related to the openness of living systems (see, e.g., Bünning 1949, p. 79). Living systems are open toward other systems; they are open toward the whole of nature. As a consequence, they can be influenced in many ways by the context around them. Since the context changes continually, the interactions with the system change also. As these interactions affect the connection of F and G, deviations from the regularity occur and repeatability of experiments becomes impossible in these instances. It is well known that the aim of scientific experimentation is to keep constant those factors that have the greatest influence on the regularity under consideration. In this way one can often obtain repeatable results. One has to keep in mind, however, that each time one repeats an experiment the global (cosmic) context is different and one can have no assurance whether this difference will affect the regularity to be studied. Thus, in principle, we have no guarantee for lawfulness and repeatability. In fact, we know from biological research many failures of repeatability. We also know many successes. They may be attributed to a relative autonomy of biological systems, which leads to a certain stability in spite of the openness (see, e.g., Thom 1975; Varela 1979).

If the changing influences on a biological system remain minimal, they are called noise. Such noise may still permit the application of probabilistic laws: F is not always followed by G, but in most cases. This approach is often taken in biology. Since it is statistical, it requires a large sample of identical or at least strictly comparable units. Obviously this requirement leads to another difficulty whenever the sample of identical or strictly comparable units is too small. Since biological units are remarkably variable – variability is a basic feature of living systems –, probabilistic prediction also faces fundamental problems because it may be difficult or impossible to determine what are comparable units.

To 2. Successful prediction of future events depends not only on general propositions such as laws, but also on a knowledge of what are called the initial conditions. Unless we can determine that F is actually the case, the law "when F, then G" cannot be applied. Kochanski (1973) pointed out various problems with regard to the complete and exact description of initial conditions [for details and examples see Kochanski (1973), pp. 45–50].

## 2.4 Understanding

Many scientists equate understanding with explanation. Whatever is explained is also understood (see, e.g., Simon 1971, p. 40). There are, however, other meanings of understanding (e.g., Bischof 1977) some of which are rather broad and transcend the limits of scientific explanation (e.g., McClintock, quoted by Keller 1983, p. 201). As we have seen, scientific explanation may depend on covering laws. Since covering laws as laws represent only the lawful aspect of nature, the chaotic or capricious aspect is excluded. Hence, a full insight is not possible as long as we limit ourselves to a scientific approach in terms of lawfulness and explanations according to the covering law model. If a full insight is considered to be understanding, then understanding is indeed far beyond explanation for it is not limited to the orderly aspect of nature. Understanding in this sense could be achieved, if at all, only through the intuitive mind that has reached complete harmony with nature. Whether it still makes sense to refer to understanding at this point or whether one should speak only of harmony is a matter of words. In fact *speaking* of harmony would also be misleading because it is *being* in harmony that matters.

In this book *Tantra – The Supreme Understanding*, Rajneesh (1975b, p. 232) contrasts knowledge and understanding. "Knowledge is always either this or that. Understanding is neither." Thus, a sage, "a man of real understanding is neither good nor bad, he understands both. And in that very understanding he transcends both" (Rajneesh 1975b, p. 233). This transcendence is in harmony with the world. Therefore, the sage "knows no trembling, no fear, and he never goes to pray to God to protect him; he is protected ... Understanding is his protection ... He lives a loose and natural, simple life, he has no predetermined concepts. And he is unpredictable" (Rajneesh 1975b, p. 233). Castaneda (1972, 1974, etc.), in his writings on Don Juan, an enlightened Yaki Indian of Mexico, comes to similar conclusions, although they are formulated differently. Don Juan makes the distinction between "looking" and "seeing." "Looking" is our ordinary experience of the world in terms of familiar categories. "Seeing" is an intuitive understanding which requires a total transformation of our ordinary existence into that of the sage who is in harmony with nature.

Some readers might object that such statements, which are founded on intuitive experience instead of objective knowledge, are too imprecise and too irrational to be taken seriously. However, on what grounds can one reject imprecision and irrationality as unnatural? What is the precise and rational foundation of the ideal of exactness and rationality? If the scientist has the privilege of believing in the lawfulness of nature, why should a sage not believe in the spontaneity of nature? Actually, a sage does not believe in spontaneity, he experiences spontaneity and thus becomes one with nature. How can anyone who has not reached this level of insight and harmony discard all this as nonsense? To do so would be as ridiculous and unwise as a colorblind person denying that others experience visions of colors. Of course, one may grant that the colorblind person has the right to be sceptical about the experience of those who claim to see colors. However, what is the use of such scepticism? It seems that the colorblind person really has no reason to be proud of his or her scepticism. Scepticism, although an important virtue, can also become a hindrance. It is often said that a virtue of a good scientist is complete openness, the ability and the drive to listen to nature patiently and to accept her answers willingly, even if they are contrary to expectations. This virtue may be practiced not only toward nature, but also toward oneself including one's scientific methodology. Thus, if we remain open to see limitations of scientific methodology and science, we may also be open to consider that there might be experiences beyond the realm of science. We cannot be as arrogant as to assume that our own experience is the measure of everything. We know from history that even within the realm of science, generations before our time could not imagine what modern science has accomplished. Scientists who took their own limited experience of their time as the measure of all things, in retrospect appear indeed ridiculous and shortsighted. Therefore, whether as scientists or as human beings, it is useful to remember Dubos' (1961) warnings: "there exist in heaven and earth more things than appears in present day scientific philosophies. To acknowledge ignorance of these important matters is, in my opinion, not a retreat from reason but instead the most constructive way to broaden and deepen the scientific approach to the understanding of life" (see also Russell 1975; Wilbur 1984).

## 2.5 Summary

*Laws.* A natural law may be defined as a regularity of events or characteristics. Thus it refers to an invariant pattern. Deterministic (= universal) laws are contrasted with probabilistic laws. A deterministic law is a statement implying that whenever F, then G (i.e., whenever conditions F occur, then invariably conditions G occur also). It is not always easy to distinguish laws from accidental generalizations which are not laws. For example, the accidental generalization that "all apples in this basket are red" is not a law, yet it satisfies the above definition of a deterministic law. However, if a deterministic law is defined as a statement saying that if F were the case, then G would have occurred, the accidental generalization is excluded from the category of laws. Nevertheless, the debate on what is the most adequate criterion for the definition of laws continues. Some philosophers think that the integration of laws into theories may be the best indicator of their lawfulness.

A probabilistic law is a statement that says when F, then probably G; or more appropriately: if F were the case, then probably G would occur. Probabilistic as well as deterministic laws may be either coexistence laws in which F and G are contemporaneous or process laws in which G follows F in time. Phylogenetic cross-section laws are examples of the former, whereas causal laws belong to the latter category. A further distinction may be made between laws comparable to axioms and empirical laws resulting from limited inductive inference.

All laws remain open to doubt. Even the question whether laws exist, i.e., whether nature is lawful, can be answered in positive or negative ways. According to the positive answer we are actually discovering the laws and order of nature. According to the negative answer we are inventing laws. These inventions, which are projections into nature, may, however, correspond partially with nature just as the concept of a circle corresponds with many structures in nature that approximate circles yet are not absolutely circular. The idea that we are inventing laws explains the success as well as the failure and limitations of laws.

In Western culture law and order has been of paramount importance. This may be related to its Judeo-Christian background in which the law was God-given originally. Later God was eliminated by many philosophers and scientists, but the law remained. In some other cultures and religions lawfulness has been of no or little importance as compared to Western culture. The present counterculture also tends to question the validity of law and order. Depending on the definition of law and order on one side, and chaos on the other side, the difference between these two sides need not be as fundamental as the words suggest. Similarly, the gap between Western culture and its counterculture is not unbridgeable.

60

*Explanation.* Scientific explanation requires explanatory relevance and testability. As these two requirements are met an explanation is possible if the phenomenon to be explained can be derived from law(s) and particular circumstances. Auxiliary assumptions and other background knowledge may also be necessary. The laws are called covering laws and consequently this model of explanation is referred to as the covering-law model of explanation. In the deductive-nomological explanation the covering laws are deterministic laws. Hence the fact (or law) to be explained follows from the covering law(s) and the particular circumstances with logical force. In the probabilistic explanation the covering law(s) are probabilistic laws and therefore the explanation does not carry logical force but only a certain probability.

It has been much debated whether the covering-law model of explanation is the only kind of scientific explanation. Narrative explanations have been proposed as an alternative, especially for historical disciplines in which it may be difficult or impossible to find relevant covering laws. Woodger looked upon explanation in a rather broad sense as analysis and relating. Jeuken uses the notion of explanation as a synonym of "clarification."

*Prediction.* Prediction in the strict sense is the forecasting of future events, whereas prediction in the wide sense is the forecasting of any unknown event. It has been much debated to what extent the pattern of prediction is symmetrical to the covering-law model of explanation. In both cases the premises consist of law(s) and particular circumstances. Some authors place enormous emphasis on prediction and predictive power and go so far as to suggest that general scientific propositions should be defined in terms of predictive power, which means that only a proposition with predictive power would be scientific. One has to keep in mind, however, that hypotheses and theories that are considered to be false can yield correct predictions and that propositions which are thought to be correct may fail to give correct predictions when coupled with wrong auxiliary assumptions. According to Kochanski (1973) limitations of predictability are related to difficulties concerning general laws, theories, and initial conditions.

*Understanding.* Many scientists equate understanding with scientific explanation. Hence, whatever has been explained is also understood. There are, however, other meanings of understanding some of which transcend the limits of scientific explanation. They may involve an intuitive insight into nature by being in a state of harmony with nature. Not seldom this way of understanding is ridiculed by scientists, especially by those who lack the intuitive insights. But how can someone ridicule what (s)he does not understand? It is just like a colorblind person ridiculing those who claim to see colors. It seems more appropriate for scientists to remain open-minded toward that which they do not understand. Both scientists and humankind could benefit greatly from such an attitude.

# 3 Facts

"Basic judgements (data, facts) are never free from *theoretical* implications" (Mohr 1977, p. 43)
"There is a sense, then, in which seeing is a 'theory-laden' undertaking" (Hanson 1958, p. 19)
"There is no such thing as an isolated fact" (DeDuve 1984, p. 17)

## 3.1 Introduction

Facts (= data) are also referred to as singular propositions, factual propositions, statements of facts, basic statements, or basic judgements (Mohr 1977, p. 42). They are often contrasted with general propositions such as theories and laws. Whereas general propositions always remain questionable to at least some extent, facts are often regarded as the hard and solid core upon which general propositions are erected and validated. In this view, facts are unyielding: we have to accept them even if they are unpleasant and contrary to our past experience, our expectations, and our most cherished beliefs. In this context, scientists often refer to "hard facts" or to "hard cold facts" (see, e.g., Hirsch 1975) in order to stress the absolute nature of facts to which we must submit unquestioningly. Such an attitude has often been considered a prerequisite for the scientific enterprise.

Although facts have to be taken seriously and should not be brushed away light-heartedly when we do not like them, the situation concerning facts is quite complicated. As we have seen in Chap. 1, facts "are never free from theoretical implications" (Mohr 1977, p. 43) and therefore they also contain an element of uncertainty and are subject to doubt. Thus, the absolutist view of facts sketched above is not tenable in such an extreme form. Nonetheless, facts are a cornerstone of science because they constitute the empirical basis of the scientific approach.

How, then, do we define a 'fact'? I think there is a tendency among scientists to consider a fact as a particularity. Since in science only that counts which is communicable in terms of a language, a fact to a scientist is a *proposition* (statement) of a particularity, i.e., a particular or a singular proposition such as, for example, the statement: "this leaf is green and red." There are also scientists who use the term 'fact' in a wider sense as all those propositions that are generally accepted by the community. If facts are defined in this way, then even some generalizations may count as facts. Some biologists who subscribe to this view refer to organic evolution as a fact. I prefer to use the term 'fact' in its narrow sense, i.e., as a singular proposition. The question that I want to discuss now is how singular propositions are related to reality. In this regard very divergent views exist depending on the philosophical position. I shall present two of those views. The first, I suspect, is of little or no use to the scientist, whereas the second applies to the scientific enterprise. Both views shall be presented in the form of more specific definitions of 'fact'.

## 3.2  First Definition of 'Fact'

A fact is defined as a description (proposition) of a particularity that is real. Following Woodger (1967, p. 133) I use "the term real for anything which may be said to exist" (see also Chap. 4, footnote 1). One might add that real is that which exists independently of us observing it, i.e., that which does not only exist for us, but which is there regardless of whether we observe it or not. This definition of 'fact' is based on the view that we observe nature as she is. Many arguments can be produced against this naive realism. For example, according to J. von Uexküll's *Umweltlehre*, which I mentioned already in Chap. 1, our experience of the world is dependent on our organization which allows us to perceive at best one aspect of the world. In this context, it is important to make a distinction between reality and an aspect of reality. An aspect of reality represents reality partially and therefore is not totally unreal, but it is not a representation of reality itself in a total sense, and hence it is not real according to the above definition of "real."

In the light of von Uexküll's doctrine one could ask the question as *to what extent* facts are a representation of reality, which leads to the second definition of 'fact'. Before I consider that definition, I have to acknowledge that there are also philosophers who deny any relationship between facts and reality [see, e.g., certain phenomenologists such as Ernst Mach whose position is critically discussed by Woodger (1967)]. As a scientist, one is tempted to be rather impatient with adherents to this extreme view. For example, when someone denies or questions whether there is a wall in front of him or her, one could simply suggest running against it and soon it will become all too obvious that this wall is a hard fact. Nonetheless, the situation is not necessarily as clear-cut as this. Changes of our state of consciousness may lead to "alternate realities" (see, e.g., Tart 1969; LeShan 1976). This means that in altered states of consciousness our environment may be perceived and experienced quite differently from our ordinary state (for an example see the description of fire-walkers in Chap. 6).

## 3.3 Second Definition of 'Fact'

A fact is defined as a proposition of a particularity that is an objective datum of perception. In other words: a fact is a proposition about an objective observation. When scientists refer to observation, they tend to use the term in the sense of fact as defined here, i.e., an observation is meant to be objective and represented in terms of a language as a proposition.

Two concepts are crucial for this definition of 'fact': perception and objectivity. I shall first deal with the former. Perception, according to Woodger (1967, chapter II), is not the same as sensing. The latter concerns what Woodger calls the primary realm. This "is the realm which 'contains' colored patches, smells, feels, sounds, pressures, temperatures, etc. It is the realm of bare awareness – a realm about which it is so difficult to speak and about which we know so little, because it is not the world of daily life" (Woodger 1967, p. 133–134). It is indeed paradoxical and misleading to speak about this primary realm, because in referring, for example, to colors, it is not the talked-about colors that are meant, but "the colors as immediately experienced or sensed, before they become objects of thought" (Woodger 1967, p. 134). Thus direct experience, what in German is called *Erleben*, occurs in the primary realm, the realm of sensing. Perception, according to Woodger, occurs when thought comes in and thus the primary realm is interpreted through the use of concepts.

It has been debated whether perception without concepts is possible. This is not surprising because the difference between sensing and perceiving is fuzzy and the two concepts are not always clearly differentiated. In the present context it is irrelevant whether perception without concepts is possible, because facts, according to our definition, are propositions of perceptions and are thus bound to be conceptual. The following conclusion is therefore unavoidable: the perception of a fact, or at least the linguistic description of it, depends on the concepts used. This means that the concepts used influence our facts. Since concepts – as we shall see in the next chapter – are abstractions from reality, facts which are described conceptually cannot be real in the total sense as required for the first definition of 'fact.' Furthermore, since the concepts we use reflect on the one hand our cultural tradition and on the other hand the theoretical framework of science, facts also mirror our culture and are theory-laden. Fleck (1979, 1981) presented many examples from medicine that show how the description of facts is influenced by the cultural context. Many other studies in the history and sociology of science have also emphasized the culture-dependence of facts. The theory-ladenness of facts has become recognized by many philosophers of science (see, e.g., Hanson 1958; Hesse 1970, 1974; Shimony 1977). This recognition is of fundamental importance as far as the understanding of facts and the

64

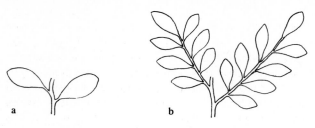

**Fig. 3.1a,b.** Outlines of portions of two plants (such as *Esenbeckia pilocarpoides* H.B.K. or certain *Citrus* species (**a**) and *Murraya exocita* L. (**b**); buds in the axils of the lateral structures are not indicated)

workings of science are concerned. The following botanical example shall illustrate the theory-ladenness, i.e., the theory-dependence of facts (Fig. 3.1). The question is: what are the plant structures that are diagrammatically represented in Fig. 3.1a and b? According to the generally accepted theory of plant morphology, which is often called the classical theory, Fig. 3.1a shows a portion of a stem with two simple leaves, whereas Fig. 3.1b presents a stem with two compound leaves. Thus, both figures illustrate two leaves on a stem. That is the fact according to the conceptual framework of the classical theory. If we consider now Arber's (1950) partial shoot theory of the leaf which is not generally accepted, but favorably looked upon by some modern plant morphologists including the present author, both Fig. 3.1a and b show a stem with two partial shoots. The difference between the two diagrams is that the partial shoots of b come closer to a whole shoot than those of a. This example shows how two different theories lead to the description of quite different facts. One might argue now that since the classical theory is the generally accepted one, the first fact is the real fact. However, the concept "real" is out of place in this context. As we know, theories are not absolute truths, but remain questionable. We cannot tell at the present time whether the classical theory will still be generally accepted in 50 years, or whether it will have been replaced by another theory. However, I do not have to speculate about the future of theories in plant morphology to support the main argument under discussion. Theories are susceptible to modification and replacement. Since facts depend on theories – or in the absence of theories, on concepts or general viewpoints that may also change – they cannot be absolute and final, but are equally subject to change. This, by the way, is another argument against a purely inductivist scientific methodology in the narrow sense: it is illusory to suppose that one may first collect all the facts and then gradually build up the scientific edifice from this foundation of facts (cf. Chap. 1).

Instead of being dependent on well formulated theories or models, facts may be influenced by concepts or points of view. An example is the description of a tree as an individual organism or as a metapopulation. The first description is based on the view that whatever develops directly from a seed and remains spatially continuous represents an individual seed plant. The second description reflects the view that elementary units of plants such as, for example, buds, may be considered individual plants. Hence, a tree that is normally described as one individual plant is a metapopulation of plants. In contrast to a population, a metapopulation is "an aggregation of parts that comprise, or are derived from, a single genetic individual" (White 1979, p. 110).

Since facts are dependent on concepts, theories, viewpoints, and the cultural context, one cannot draw a sharp line between a fact and the theoretical interpretation of a fact. In a sense each fact is already interpreted. There is no absolute neutrality of facts. Nonetheless, certain facts appear "harder" to us than others. This may be so because they depend on assumptions that are so generally accepted that we may no longer be aware of them. Once these assumptions are questioned or changed, we notice the relativity of the facts that are altered by this change (see, e.g., Fleck 1979, 1981).

## 3.4  Objectivity

It is important to distinguish the following two definitions of 'objective':

(1)  objective is what is real
(2)  objective is what is shared.

The first definition is often implied by scientists as well as philosophers. For example, Monod (1970) seems to equate objective with real and true. If this meaning of objective is implied with regard to our second definition of 'fact,' then a fact becomes a real datum of perception or a real observation, and if a real observation is meant to be an observation of something that exists independently of us observing it, then the equation of objective with real makes us fall back onto our first definition of fact.

According to the second definition of objectivity, a datum of perception is objective if it is shared by different observers. Hence objectivity in this sense means inter-subjectivity, i.e., different subjects (observers) share the same experience or perception. This kind of objective perception contrasts with the subjective perception that is not shared by different observers, but is a unique perception of one particular subject. Subjective experiences or perceptions are private, whereas objective ones are public (see, e.g., Huxley 1963). It is generally said that science has to limit itself to public experiences, whereas in the arts and humanities subjective experiences are explored as well. Thus, science in accepting this limitation remains objective according to our second definition, whereas this need not be the case for the arts (including music) and the humanities.

Insisting that science has to be objective (in the sense of inter-subjective) does not mean that it cannot investigate subjective experiences. But in dealing with subjective experiences, science has to use an objective approach, i.e., different subjects have to share the same perceptions with regard to the subjective experiences under investigation.

Many scientists seem to think that an objective experience (in the sense of inter-subjectivity) must be true or real. This is a dangerous fallacy because sharing an experience does not necessarily mean that it must be real. Mass deceptions are shared experiences, yet they do not seem to be true or real, i.e., what is experienced does not seem to exist objectively in the sense of our first definition of objectivity (see Fig. 3.2 and Robinson 1972).

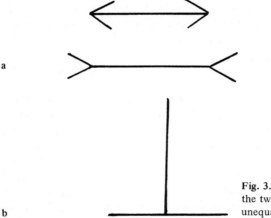

a

b

**Fig. 3.2a,b.** Most observers perceive the two lines in a and b as being of unequal length

It is not even necessary to refer to mass deception. It would be sufficient to look at the history of biology which provides many examples of experiences shared by the scientific community which were later shown to be unwarranted. The masses can be easily misled in many ways, and the masses of scientists are not immune to error.

As objectivity (= inter-subjectivity) is often falsely equated to truth or reality, so subjectivity is often taken as false or untrue. However, just because an experience is subjective (i.e., not shared by others) it does not mean that it must be false. For example, the subjective experiences of at least some artists may be closer to the truth or reality than many of the objective experiences of scientists. Even in science new insights may originate from subjective visions that only subsequently face rigorous and objective analysis.

Hence, the fact that something at a certain time is perceived only by one or a few individuals is not a criterion for its existence or nonexistence (see also Rose 1983).

This brings us to another important consideration. What at one time or one place may be a subjective experience, may become objective at a later time or another place. It is known that at least some experiences change as a culture develops. We have witnessed a rather rapid change of experiences during the last decades in Western society. Stent (1969, 1978) gives us a vivid description of some of these developments. We also know that at least some experiences may change from one culture to another. Thus what is considered subjective or objective (in the sense of inter-subjectivity) is relative. It depends on the state of evolution of society. It depends on which conceptual framework is adopted for the description of our facts and which state of consciousness is favored as a basis for experiencing and perceiving the world. For example, if the majority of us in Western society developed the state of consciousness of the fire-walkers mentioned in Chap. 6, it would become an objective fact that burning coal does not burn our feet. At the present time this is still subjective because it is experienced by very few.

Stent (1969, 1978) thinks that our Western society has been undergoing a profound transformation and is "on the road to Polynesia." If we ever arrive in this paradise or "Golden Age" our perception of the world will change drastically and many experiences that today are still subjective will become objective. Thus, if some people in the "Golden Age" still feel like doing science, many of their objective facts might be quite different from our accepted facts today.

Quite apart from speculations about the "Coming of the Golden Age" (Stent 1969), we cannot but conclude that facts (according to our second definition) are relative. This relativity becomes especially obvious in periods of transition from one culture to another or from one paradigm to another. In such periods certain subjective experiences gradually become objective as more and more people begin to share them. Because of the gradual transition it may become impossible to tell at a certain stage of development whether a certain experience is still subjective or already objective. Those to whom the experience has been familiar for a long time and who have been living in a subculture in which this kind of experience is common will not hesitate to call it objective, whereas members of another subculture to whom the experience is new will be reluctant to accept it as an objective experience. This discrepancy explains why in our present age of cultural transformation we are engaged in many controversies over what is objective and what is scientific. Consider, for example, the differing views held on the status of astrology or parapsychology that according to some are objective and scientific and in the opinion of others constitute merely subjective pseudo-science.

In conclusion, we may state that there is a double aspect to the relativity of facts. First, facts are relative to the culture. Where cultural (and scientific) traditions are well established, it may be quite clear what is a fact and what is a subjective experience. Second, in a period of cultural (and scientific) transition, facts are also relative to the standpoint of an observer within the culture (or scientific enterprise).

## 3.5 Facts and Reality

Instead of using the term 'fact', scientists often refer to 'data.' The literal meaning of 'data' suggests that they are given (cf. the French term 'donnée' which also means given). In other words: the term 'data', by its etymological derivation, reinforces the idea that what is described is simply "out there," a particularity of reality as it exists independently of our observing it (cf. Chap. 4, footnote 1).

The term 'observation,' if we take its literal meaning, may suggest at least that what is "out there" is being observed so that an interaction between the observed and the observer may be visualized. If this interaction influences the observed, as may be the case in microphysics, biology, psychology, and sociology, then we can no longer assume that the observed is real in the sense that it exists independently of our observing it. In this case the observer is an integral part of the observed. In such a unity of the observer and the observed, or the subject and the object, objectivity according to our first definition becomes illusory (see, e.g., Bateson 1979).

The term 'fact,' which is etymologically derived from the latin word 'factum,' indicates that what is described is made by us (cf. the French word 'fait' which means 'made'). Thus, according to the etymology, facts are not "out there," but they are our doing (cf. the German word for fact, 'Tatsache', which etymologically means something that results from an act or a deed of the observer). In terms of the preceding discussion this means that facts are produced by our act of perception in a certain state of consciousness, and influenced by the conceptual framework used for description. We may ask now whether under these circumstances there are indeed any connections between our facts and reality. The answers given by different people are very divergent. It has been said that because facts are made by us anything can be accepted as factual provided one allows for enough changes in the conceptual framework and the states of consciousness. This extreme point of view has been taken, for example, by Pearce (1973). I find it difficult to accept this position. For example, the fact 'there is one cat in this room' may be replaced by other facts if the state of consciousness and the conceptual framework are different. However, I do not see how it could be replaced, for example, by the fact 'there are two cats in this room.' Hence, I think, that there are certain constraints given through reality, and for this reason I do not think that "anything goes." Many things may go, but not anything. And since not "anything goes," we cannot conclude that facts are solely products of the observer and totally arbitrary as far as reality is concerned. I agree with Woodger (1967, p. 135) who wrote: "The objects of knowledge are *in a sense* ... creations of the intellectual activity" (and the state of consciousness, I would add) "just as much as the objects of art are creations of the aesthetic activity. But in neither case are they *free* creations" (for a more detailed discussion of this point and examples see the following chapter on Concepts and Classification).

To avoid misunderstandings, I want to distinguish now two kinds of relativism: on the one hand the extreme relativism of Pearce (1973) and others according to which "anything goes," and on the other hand a moderate relativism which I have been advocating in the preceding discussion. Moderate relativism does not claim that facts are real; it emphasizes that facts may depend on the observer and therefore may be relative. However, it concedes that reality enters at least to some extent into the facts. The question to what extent reality is represented by facts is debatable and I do not see how it can be answered in any general way, if at all. One really needs to know reality in order to judge to what extent a fact represents it.

Another argument against the idea that facts are solely our creations may be added. If facts did not mirror reality at least to some extent, how then would it be possible to succeed in science so that we can make predictions about the future? Furthermore, how would it be possible for us to survive in the world if all our observations on which we depend to orient ourselves were fictitious? Riedl (1980) tells the revealing story of academics who were debating the reality of facts; when approached by a herd of rhinoceri they immediately ran for safety. After all, rhinoceri are dangerous. However, we have to remind Riedl that perhaps rhinoceri are not always dangerous and perhaps not to everybody. The same could be said of lions. An Indian artist portrayed a sage sitting peacefully with two lions under a tree.

Now some comments are necessary on the singularity or particularity of facts. Since facts do not refer to the whole, but only to particularities, they cut out segments from the unity of the world. Thus the observer does not only perceive the facts through the "glasses" of his perception, but the delimitation of the facts from the whole context of reality is also the creation of the observer. Again we may ask whether this creation is totally arbitrary or whether it is constrained by reality. I would think that it is not totally arbitrary, but may reflect reality to some extent. Although reality appears to be a unity, it does not seem to be uniformly homogeneous. It appears to be patterned in the sense of being a "patterned continuum" (Weiss 1973).

Particularities thus reflect regions of reality that are partially (but not totally) isolated from the surrounding regions (for a more detailed discussion see the following chapter on Concepts and Classification).

In conclusion, it may be said that facts do not represent reality as it is independently of our observing it. First, our perception projects our state of consciousness and our conceptual framework onto the facts we describe. Second, our perception isolates particularities as entities from the patterned continuum of the unified world. From this point of view, we may state that facts are not real. However, they do not seem to be totally fictitious either, but may represent reality to some extent.

### 3.6 Experience

As the terms 'fact,' 'objectivity,' and 'reality' have different meanings, so, too, the term 'experience' is used in different ways (see, e.g., Le Shan and Margenau 1982). In the following, I shall describe some of the meanings of the verb 'to experience.' First of all, to scientists experiencing can be perceiving. In that sense particularities are experienced and these experiences are described as facts. Second, experiencing can also be understood in a more direct way, namely as sensing as defined by Woodger (1967, p. 134). This is an experience of what Woodger (1967, p. 134) calls the primary realm of the world. In contrast to the first kind of experience (perception), here thought is not projected into this realm. Therefore, this experience is an immediate or direct experience of the world which has not yet been divided up by discursive thought. Consequently, facts as particularities do not exist. Experience in this direct sense does not constitute knowledge because knowledge requires thought. Only experience according to the first definition is knowledge. Naturally, a clear-cut distinction between the two kinds of experience is not always possible in our daily activities. Especially the direct or immediate experience is very difficult to obtain for most of us. We are habitually too conditioned by discursive thought to be able to experience directly without the intervention of thought. Thus in many instances only an approximation of direct experience may be possible.

Finally a third kind of experiencing that goes beyond sensing may be distinguished, but this way of experiencing is possible only for an enlightened person who has been totally aware of his or her integration with the world. At this stage there is no longer "any distinction between *his* [her] self and other selves, the subject and the external objects. And this is the experiencing of the Absolute as understood in Zen Buddhism" (Izutsu 1971, p. 505). It coincides with an enlightened state of being. There is neither subject nor object of experience.

## 3.7 Summary

*Introduction.* Facts (= data) are also referred to as singular propositions, factual propositions, statements of facts, basic statements, or basic judgements. Like generalizations, they are subject to doubt. Two definitions are discussed:

*First definition of 'fact.'* According to this definition, a fact is a proposition of a real particularity. This definition is not useful in science because it would require an independent yardstick of reality. Furthermore, a number of arguments show that a fact cannot be real in the sense of existing independently of our observation.

*Second definition of 'fact.'* A fact is a proposition of an objective sense datum. This implies that a fact is dependent on the perception of the observer which again depends on his state of consciousness and the concepts chosen for the description of the fact. Inasmuch as the concepts used form an integral part of a theory, facts are theory-laden. Inasmuch as the concepts used reflect the cultural tradition in which they are rooted, facts are dependent on the cultural background of the observer.

*Objectivity.* Objectivity can be defined in terms of reality or as inter-subjectivity. The latter definition is useful in science. It implies that only those facts are admitted that are recognized by the consensus of different observers. Such shared experiences are not necessarily closer to reality than subjective ones that are not shared. What at one time and place is subjective may eventually become objective if the culture evolves in such a way that the unique subjective experiences become the norm. In transition periods a clear-cut distinction between what is subjective or objective may not exist.

*Facts and reality.* Facts do not represent reality as it is independently of our observing it because our perception projects our state of consciousness and our concepts onto the facts we describe, and, furthermore, our perception isolates particularities as facts from the patterned (or heterogeneous) continuum of the whole world. From this point of view, facts are not real. However, they may represent reality to some extent.

*Experience.* Three meanings of "experience" are distinguished: (1) perception, (2) sensing that does not involve thought, and (3) seeing as being of an enligthened person such as a Zen master. Only experience as perception provides facts in the sense of particularities.

# 4 Concepts and Classification

"Nothing could be further from the truth than that there is complete correspondence of structure between all experience and all language, or even any limited aspects of language and experience" (Bridgman 1936, p. 23).

"Basically, everything is one" (Barbara McClintock, quoted by Keller 1983, p. 204).

## 4.1 Concepts

### 4.1.1 Introduction

Science, like philosophy, is conceptual, i.e., scientific statements are made through the use of concepts. Since concepts are so basic to any scientific statement, it is important to be aware of what a concept is and what it is not. A concept is an abstraction. What do we do when we abstract? We separate certain feature(s) from the welter of our direct (immediate) experience (Langer 1964, p. 66). The verb "to abstract" is etymologically derived from the latin 'ab-trahere' which literally means 'to draw away.' Thus, in forming a concept through abstraction we separate feature(s) and this selective separation of feature(s) constitutes the concept. For example, in forming the concept of a constellation of stars such as the Orion we select a number of stars, particularly seven principal ones in the well-known arrangement. The features that are selected in this instance are individual stars. This shows that what is selected is in nature. However, the selection is our own doing, i.e., our abstraction. It is evident that, given the whole world or even the smallest part of it, there is a multitude of possibilities for the selection of feature(s) and consequently for the formation of concepts.

Different conceptualizations of colors in different cultures present another interesting example of concept formation. According to Gleason (1961), the Shona tribe in Rhodesia and the Bassa tribe in Liberia use concepts of colors that differ remarkably from those normally used in Western culture (Fig. 4.1). Whereas in Western culture we divide up the spectrum into six entities and thus use the concepts of "violet," "blue," "green," "yellow," "orange," and "red" (Fig. 4.1a), the Bassa tribe distinguishes only two colors, "hui" and "ziza" (Fig. 4.1c), and the Shona tribe selected three areas in the spectrum of colors, namely "cips$^w$uka," "citema," and "cicena" (Fig. 4.1b). The

74

English:

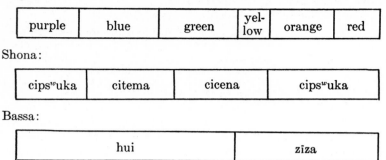

| purple | blue | green | yel-low | orange | red |

Shona:

| cips$^w$uka | citema | cicena | cips$^w$uka |

Bassa:

| hui | ziza |

Fig. 4.1. An approximate representation of the abstraction (partitioning) of colors in three different languages (From: An Introduction to Descriptive Linguistics rev by H.A. Gleason. Copyright (c) 1961 by Holt, Rinehart and Winston. Reprinted with permission of CBS College Publishing)

comparison of the conceptualizations of the Bassa tribe and Western culture is easy because the line drawn between the two colors of the Bassa tribe corresponds with the line between green and yellow of the Western scheme. Hence, one can look at the conceptualization of the Bassa tribe simply as less differentiated than that of the Western scheme. This means the Bassa tribe lumps together as one color our purple, blue, and green and does not distinguish between yellow, organe, and red (see also Gleason 1961, p. 5). A comparison of the conceptualization of the Shona tribe with those of the Bassa tribe and of Western culture is more complex because not all the limits between the colors of the Shona tribe coincide with the limits drawn in the other two schemes (see Fig. 4.1). Furthermore, "citema" also includes black, and "cicena" white. "Cips$^w$uka" comprises the beginning and the end of the spectrum. From this it is evident that members of the Shona tribe give a different description of the colors of the world than we do. Their description can be considered complementary to ours because it is based on a complementary abstraction (see Sect. 4.1.3).

Another example that illustrates different ways of abstraction and concept formation is the subdivision of biology into subdisciplines. Using a taxonomic criterion one can subdivide biology into botany and zoology (and some other fields such as bacteriology). In terms of levels of organization biology consists of molecular, cellular, organismal, and environmental biology. And according to other features of living organisms one may distinguish between morphology, physiology, genetics, systematics, etc.

Separating certain features and thus forming a concept often requires the drawing of boundaries. From this point of view abstraction and concept formation are a problem of delimitation. The example of colors illustrates how different delimitations lead to different concepts.


### 4.1.2 Nominalism Versus Realism

Having pointed out how concepts are formed, we may now ask what their status is with regard to reality [1]. Two extreme views have been taken already in the middle ages (and before). The nominalists think that concepts are only names that lack any relationship to reality. Thus, as far as reality is concerned, any name is just as good (or as bad) as any other one. The conceptual realists claim, however, that at least some of the concepts are real. This means that they "exist in themselves and would exist even if there were no minds to be aware of them" (Woozley 1967). Hence, they are not the result of abstraction.

Real concepts are also called universals. The controversy between nominalists and realists reached a peak during the middle ages in the dispute over universals. It is not at all over and it seems it will never be totally finished. Among biologists we also find nominalists and realists. For example, species by some are considered just as names devoid of a representation of reality, whereas others maintain that species are real, i.e., they are natural kinds (see, e.g., Kitts and Kitts 1979). Maybe it is fair to say that most of the practicing biologists are conceptual realists or have a tendency toward this position. Thus, to them entities such as "genes," "cells," "organs," "organisms," and "species" are real (i.e., natural kinds), whereas higher taxonomic categories such as families, orders, or phyla are often considered to be unreal or only partially real. There are also authors who take a position somewhat intermediate between extreme nominalism and a realism that postulates natural kinds defined by the possession of common (essential) properties [see, e.g., Dupré (1981) who defends a "promiscuous realism" according to which species lack essences because they are seen as peaks in a continuum]; although this view may come closer to nature than the postulate of natural kinds (essences), it is questionable whether it applies to "ochlospecies with a mainly unresolvable, reticulate variation pattern" (Whitmore 1976, p. 25).

---

[1] As pointed out by Bohm (Bohm in Wilber 1982, p. 53), the term 'reality' "is based on the word *res*, meaning thing, and the thing is what is known." Furthermore, "*res* is based on the word *rere*, mening to think, and the thing is what you can think about, essentially. So reality is just what man can know" (Bohm in Wilber 1982, p. 53). In contrast to this literal definition of reality, in common usage reality tends to be equated with truth, i.e., "that which is" (Bohm in Wilber 1982, p. 64). I am using the term 'reality' in the latter sense.

### 4.1.3 Degree of Adequacy of Concepts

I think that neither realism nor nominalism are tenable in the extreme forms as described above. Since concepts are abstractions *from reality*, they are not totally fictitious; they do represent at least certain features of reality. Woodger (1967, p. 153) uses the analogy of a map to illustrate the relation between concepts and reality. A map represents a certain region of the world. If it is a good map, it allows us to find our way around. How could this be possible if there were no correspondence whatsoever between the map and the region of the world it represents? The better the map, the greater is the correspondence with reality; the greater the correspondence, the higher the degree of adequacy. With regard to concepts this means that the more concepts correspond with reality, the more adequate they are.

Woodger's map analogy also shows that even the most adequate concepts can not represent reality completely. Even a very good map can at best be an approximation to reality. For example, what in the map is represented as a straight line is not a completely straight line in nature. In this sense a map and also a concept are simplifications of nature. Furthermore, a map presents only some features of a region of the world such as, for example, distances. It does not represent that region as it is in its total existence "out there." However, it can incorporate more and more features and thus become more complete and more adequate. If features are selected that are highly correlated with others, the degree of adequacy is also greatly enhanced. However, since the correlation is never complete, no single map can represent all aspects of a region. For example, a geomorphological map may be partly (or even totally?) correlated with a geological map of the same region, but a political or economic map will show different aspects. Maps that show different aspects and are thus complementary to each other, may be equally adequate. In this situation it is not useful to ask which one is better.

In his description of the map analogy, Woodger also points out that not all concepts are comparable with maps since all maps have "perceptual exemplification," but not all have concepts. "Perceptual exemplification" means that the concept refers to something that can be perceived as maps refer to regions of the world that can be perceived at least in principle. An example of a concept without perceptual exemplification is the concept "two" which cannot be perceived. We can perceive two things, but not the concept "two." Thus, whereas all objects (of our perception and thought) are concepts, not all concepts are objects (i.e., thought-objects) with a perceptual exemplification. However, all concepts have in common that they are "representative *of*, i.e., thoughts *of*, something else" (Woodger 1967, p. 155), either an object or something which cannot be represented pictorially.

Biological examples that illustrate the preceding discussion abound. Practicing taxonomists often distinguish between "good" and "bad" species. "Good" species are concepts with a high degree of adequacy, whereas "bad" species are comparable to a poor map. Taxonomists also make a distinction between artificial and natural classes. Quite often it is implied that artificial classes are concepts devoid of any real basis, whereas natural classes are looked upon as real. I hope that the preceding discussion has shwon that neither is the case. Even the most artificial class that is based on only one feature such as "being aquatic" in the case of "aquatic plants," presents an aspect, although an extremely limited aspect, of nature. For this reason even a so-called artificial concept may be useful for our orientation in the world. And the most natural class such as "orchids," which is based on a highly correlated assemblage of features, is still an abstraction and in this sense not real (as defined above).

Biologists also use different kinds of species concepts (see Sect. 4.2.1) and continue debating which is the best one. Although it is possible that some of these may be less adequate than others, they can also be considered complementary to each other. For example, the holomorphological species concept that is based on overall similarity may be complementary to the biological concept that is based on reproductive isolation or related phenomena. The two are not necessarily correlated with each other, although this may often be the case.

### 4.1.4 Limits of Concepts

Philosophers of science make a distinction between the intension and the extension of a concept. The intension (= connotation) of a concept is the defining set of properties, whereas its extension (= denotation) is the domain of applicability. There may be concepts that have an intension, but no extension. For example, if soul is defined as something immaterial, this concept of soul has no extension to a materialist; to him or her it is an empty set, i.e., there are no objects in the world that satisfy it.

If concepts that have an extension are applied beyond their domain of applicability, they become meaningless. Woodger's map analogy may illustrate also this point. A map refers to a certain limited region of the world. To apply it beyond its limits is meaningless and creates confusion. In biology, it is not infrequent that concepts are applied beyond their limits. An example is a certain approach toward individuals that are intermediate between species. If individual x is such an intermediate between species A and B, it does not make sense to apply the mutually exclusive concepts of species A and B to x. Thus, as already pointed out in the Introduction, it is a misguided question to ask whether x belongs to species A or B because it does not fall into the domain of either one of the two species. Many examples of such cases have become known (see, e.g., van Valen 1976 on oaks). A failure to recognize the limits of applicability of concepts is characteristic of many fruitless debates in both science and everyday life.

### 4.1.5 Classification of Concepts

Concepts have been differently classified. I shall present only two classifications. One distinguishes qualitative, comparative (= semiquantitative), and quantitative concepts. If we take coloration to illustrate these three kinds of concepts, "red" is an example of a qualitative concept, "more red" (than yellow) is an example of a comparative concept, and the tone of color in terms of wavelength is an example of a quantitative concept. Quantitative concepts are especially useful in science because of their accuracy and great empirical content. They provide good tests for hypotheses and theories. Qualitative concepts are often treated as universals or natural kinds. In that case they become by necessity mutually exclusive. Thus if, for example, "red" is considered a reality, then it follows that something is either red or it is not. In terms of essences, i.e., the philosophy of essentialism, this means that something is either essentially red or not. If it is essentially red, it manifests the essence, i.e., the reality, of red. Essentialism, which is related to conceptual realism, can be a great obstacle to scientific research [see, e.g., Popper (1962, 1966), for a general criticism, Hull (1965), Sattler (1966, 1974a,b), and Mayr (1976a, 1982) for biological case studies that show the negative influence of essentialism on biology].

Another classification of concepts distinguishes between individual concepts, class concepts, relation concepts, and quantitative concepts (see Bunge 1967, p. 60). An example of an individual concept is "Charles Darwin"; "biologist" is a class concept; "between" is a relation concept. Individual concepts are often mistaken for real entities of nature that are supposed to exist independently of our observation and conceptualization. It seems paradoxical to consider Charles Darwin a totally isolated entity of nature. He himself pointed out the continuity of evolution. "Charles Darwin" probably is a concept with a high degree of adequacy, but it still is an abstraction. Our constant use of language reinforces the belief that individuals actually exist as entities independently of our conceptualization (see also below).

### 4.1.6 Term, Concept, and Reality

It is important to make a distinction between the concept and its term. The term is merely a symbol, a word, a name that consists of letters. It designates the concept and it refers to reality. The concept is an abstraction that refers to reality. In order to distinguish concepts and terms, a standard convention of philosophers that I am following in this book is to place concepts in double quotation marks and terms in single quotation marks. For example, the term 'red' designates the concept "red" which like the term refers to what is red in the real world.

A distinction between terms and concepts is necessary because the same term may be used to designate different concepts and the same concept may be represented by different terms. For example, the term 'verification' may be used to designate the concept of "verification" (= "proof") or the concept of "confirmation" (see Chap. 1). And the concept "proof" may be represented by the terms 'proof' or 'verification.' Since we have more concepts than terms, many words stand for more than one concept. Much ambiguity arises for this reason.

Many of the terms used in the preceding pages designate different concepts. 'Fact,' 'objectivity,' 'real,' 'hypothesis,' 'explanation,' 'verification,' 'falsification,' and many others have at least two meanings. In some cases one can avoid confusion by adding 'sensu stricto' (= narrow sense) or 'sensu lato' (= wide sense) behind the term. Thus, one may distinguish between 'hypothesis s. str.' and 'hypothesis s. lat.' In other cases one has to add special qualifications in order to avoid confusion. For example, one can refer to 'objectivity in the sense of inter-subjectivity.' Or one can stipulate that one uses an ambiguous term in only one specified sense. For example, I can point out that in this book I am using the term 'objective' to designate only "intersubjectivity." When I want to refer to reality I do not use the term 'objectivity,' but the term 'reality.'

The following scheme that is modified after Bunge (1967, I, p. 57) shows the relation between terms, concepts, and reality in a more comprehensive context by distinguishing three levels:

| Levels | Example |
|---|---|
| Linguistic level (terms, phrases, languages) | 'Cell' (= a symbol or word of four letters) |
| Conceptual level (concepts, propositions, theories) | "Cell" (= a concept defined by certain properties) |
| Level of reality | Cell (= problematical entity because it is questionable whether it actually exists independently of our conceptualization) |

Bunge (1967, I, p. 57) refers to the physical level instead of the level of reality. If it is contended that reality is only of a material nature, this is appropriate. I think, however, that reality is neither physical nor spiritual, but the Unnamable (see Chap. 2). I also think that there is more and more evidence from a variety of scientific disciplines and a profound experience of the world indicating that reality is not just atomistic, i.e., composed of a hierarchy of entities, but that it also is a unity (see, e.g., Bateson 1972, 1979; Capra 1975, 1982; Bohm 1971, 1980). This view does not imply that atomistic concepts are totally useless and inadequate. Atomistic concepts such as "atom," "cell," or "individual" (organism) may capture and represent the discontinuous aspect of reality. In this sense atomistic concepts are complementary to those underlining the continuous aspect of reality. If we want to make any general statement at all about the nature of reality and of life it might be best to speak of a "patterned continuum" (Weiss 1973). Since it is a continuum it forms a unity, but since it is patterned, elements can be distinguished. However, these elements are not totally discontinuous and therefore an atomistic view of reality that posits a hierarchy of entities such as elementary particles, atoms, molecules, cells, organs, and organisms does not seem to be real in the sense of existing independently of our conceptualization.

## 4.1.7 Definition

A definition describes the meaning of a term (or concept). For example, the definition of 'biology' is as follows:

$$\text{'Biology'} = \text{'Science of life'}$$

'Biology' is called the definiendum, i.e., that which is to be defined, whereas 'Science of life' is the definiens, i.e., that which defines. "To define a term is to show how to avoid it" (Quine, quoted by Hempel 1966), because the term of the definiendum should not recur in the definiens.

The maxim 'Define your terms' is a very important one, especially since many terms have more than one meaning. Unless a definition is given, confusion may arise. I think that quite a number of controversies in science are not disagreements on real issues, but are due to semantic misunderstandings and confusion. Quine's (1961) contention that real and semantic issues cannot be sharply delimited, complicates the matter.

Although definition of terms is of fundamental importance, we have to realize that not all terms can be defined. If we define one term by other terms, we may have to define these other terms again, and so on. Since there is no end to definitions, we are bound to stop at one point and leave the terms of the definiens of that last definition undefined. These terms that are no more defined are called primitive terms. We have to grasp the meaning of these primitive terms intuitively. Thus intuition comes again into play and since all defined terms are eventually based on intuition it can be said that the whole edifice of science with all its terms fundamentally rests on intuition. Anyone who claims that science can or should be totally exact should reconsider this claim in the light of primitive terms.

Naturally, one may choose one's primitive terms as one pleases. For example, one may adopt 'identity' as a primitive term or one may define it by other primitive terms. However, one term or another will have to remain undefined.

### 4.1.8 Operationism

Operationism is a philosophy concerning definition, introduced by the physicist Bridgman (1927). A clear account is also given in the excellent book on *The nature of physical theory* (Bridgeman 1936). Operationism has become popular in biology, and in many circles is imperative (for a critique see Hull 1968).

According to operationism, "we mean by any concept nothing more than a set of operations; the concept is synonymous with the corresponding set of operations" (Bridgman 1927, p. 5). In other words: the meaning of a term is determined by operations that provide a criterion for its applicability. This criterion is called an operational definition. It guarantees that terms are only applied within their domain of applicability.

As an example the following operational definition of 'length' may be given (Hempel 1966, p. 89):

'Length' = operation that involves the use of rigid measuring rods.

It is obvious that this definition of 'length' is very limited. A curved surface has no length according to this definition because the operation of applying a rigid measuring rod to a curved surface is not possible. To avoid this shortcoming one could use another operational definition of 'length' that involves the application of flexible measuring rods as the criterion of the definition. According to this definition, curved surfaces have a length, but the distance between stars has no length. At this point one could introduce yet another operational definition of length which relies on optical tests for the determination of length.

As Hempel (1966, p. 91) pointed out the insistence on operational definitions leads to a multitude of concepts. This defeats one of the principal purposes of science, namely the use of very general concepts (see Hempel

1966, p. 94). Another shortcoming of operationism is that certain phenomena may be undetectable. Hempel (1966, p. 97) discusses the assumption that all distances in the universe change steadily in such a way that they double every 24 h. This assumption implies also a doubling of measuring rods. Hence the phenomenon cannot be detected if the operational definition is used. It is meaningless to the operationalist, although it may occur.

Biologists who embrace operationism are not always fully aware of its implications. Many of them also tend to depart more or less from the original philosophy as expounded by Bridgman. For example, according to Sneath and Sokal (1973, p. 17), "operationism implies that statements and hypotheses about nature be subject to meaningful questions; that is, those that can be tested by observation and experiment."

### 4.1.9 Are There Biological Kinds of Concepts?

There is a difference between a biological concept such as "gene" or "cell" and a specifically biological *kind* of concept, i.e., a kind of concept that exists only in biology (and maybe in psychology or the social sciences), but not in the physical sciences. The question that I am posing is whether there are such biological *kinds* of concepts. Beckner (1959) discusses this question in his book on *The biological way of thought* and answers it negatively. However, he underlines that there are three kinds of concepts that, although in principle not limited to biology, are characteristic of biology in contrast to the physical sciences such as physics or chemistry where they are hardly of importance. We may call them typically biological kinds of concepts. They are the polytypic (= polythetic), functional and historical concepts. Functional and historical concepts are well known. For example, organs may be defined by their function, and "hybridization" can be visualized as a historical concept because it is defined as a certain process occurring in history.

Polytypic (= polythetic) concepts are less known and therefore will be explained in detail. They are characterized by the following three conditions (Beckner 1959, p. 22): Suppose that we have a group K with an aggregation G of properties $f_1, f_2, f_3$ ... such that

(1) each individual of K possesses a large number of the properties of G,
(2) each property of G is possessed by a large number of individuals of K,
(3) no property is possessed by all individuals of K.

If conditions 1 and 2 are fulfilled, the class is polytypic. If all three conditions are satisfied the class is fully polytypic.

As a general example I take a group K of five individuals (1–5) with an aggregation G of five properties $f_1, f_2, f_3, f_4, f_5$. I distribute the properties in such a way that the class will be fully polytypic:

| | | | | |
|---|---|---|---|---|
| 1. | | $f_2$ | $f_3$ | $f_4$ | $f_5$ |
| 2. | $f_1$ | | $f_3$ | $f_4$ | $f_5$ |
| 3. | $f_1$ | $f_2$ | | $f_4$ | $f_5$ |
| 4. | $f_1$ | $f_2$ | $f_3$ | | $f_5$ |
| 5. | $f_1$ | $f_2$ | $f_3$ | $f_4$ | |

Provided four out of five is considered a large number, then the above example represents a fully polytypic class. One point of fundamental importance that Beckner demonstrated by his analysis of fully polytypic classes is the fact that these classes can no longer be defined in the traditional sense because their members have not even a single defining property in common. The great merit of Beckner's analysis was to show that the notion of a fully polytypic class allows us to conceive of classes where the traditional approach fails.

The traditional concept of classes is called monotypic (= monothetic). Monotypic classes are defined by a set of properties or at least one single property. This concept of monotypic classes is only of limited usefulness in biology. Living systems usually are too variable to conform to fixed sets of properties. Intuitively this has been known for a long time, especially by competent systematists. They have, however, failed to point out the logical structure of polytypic concepts. When I was a student I could not understand why taxonomic classes were not well defined monotypically. Most of the definitions of such classes were followed by a list of exceptions. These exceptions, although they did not satisfy the definition, were nonetheless considered as belonging to the class. I found this situation very dissatisfying. From a logical point of view, I felt that the individuals that did not satisfy the definition should be excluded. However, intuitively such exclusion appeared to be artificial and unjustified as it was recognized by competent systematists. Beckner's (1959) book revealed the solution to this puzzle still during the time of my graduate studies, although I learned of it much later. Actually, Wittgenstein had made a similar point long before Beckner, but I was not aware of that either.

A botanical example of fully polytypic classes are the two major groups of the flowering plants, namely the monocotyledons and the dicotyledons. The dicotyledons usually are defined by the following set of properties: two cotyledons (seed leaves), leaves with netted venation, vascular bundles of the stem arranged in a cylinder, vascular cambium present, floral parts in whorls of four or five (or higher numbers, but not three). One can find exceptions to all of these properties. Thus there are, for example, dicotyledonous plants with only one cotyledon. Since these plants share the other properties listed above, they belong to the fully polytypic class of dicotyledons. It would be wrong to conclude that such classes cannot be defined as is sometimes claimed. They cannot be defined monotypically, but they can be defined in the fully polytypic sense.

I think it is obvious that the notion of polytypic classes is also of great significance in everyday life. So often we try to find traditional monotypic classes which, if imposed rigidly, almost inevitably exclude members that belong to the class in a polytypic sense. Such cases become of great importance when law and justice is concerned. Too often it may happen, for example, that someone does not qualify for social benefits because one of the defining properties of the whole set is not satisfied. In such cases we feel the injustice intuitively, but often nothing can be done when the monotypic approach is followed. Consequently there are still many victims of the notion of a monotypic class. In extreme cases, the difference between monotypic and polytypic classification may be a matter of life or death.

Although the concept of (fully) polytypic classes allows us to deal with many problems of classification much more satisfactorily than that of the traditional monotypic classes, there are certain limitations to the use of the (fully) polytypic concept. Beckner's first and second condition contain the notion of "large number." What constitutes a large number? 9 out of 10, 8 out of 10, or still 7 out of 10? We cannot draw a sharp line and therefore (fully) polytypic classes cannot be clearly delimited from a collection of individuals that no longer constitutes a class in either a monotypic or (fully) polytypic sense. How do we deal with such situations? Zadeh (1965, 1971) presented a solution by his "fuzzy set theory." According to this theory, class membership is no longer total, but a matter of degree. Any individual may be a member of a class from 0% to 100%. Being a 0% member of a class is the borderline case in which the individual does not belong any more to that class at all. A 1% member would still partially belong to the class and would be closer to the 0% member than, for example, to a 10% member.

Fuzzy set theory so far has been very little used in biology (see, e.g., Marchi and Hansell 1975; Dubois and Prade 1980; Beatty 1982). Nonetheless, I think that this theory has enormous potential in biology because it can cope successfully with fuzzy situations that are so typical of life. It also should prove very useful and adequate in psychology and the social sciences as well as in everyday life situations and in politics. Communists and capitalists are often treated like members of two monotypic classes that are defined by a set of properties. What we find in real life is a whole range of membership from 0% to 100%. Many people actually are partial members of both fuzzy sets. Much unnecessary antagonism would disappear if we just realized that we do not belong to mutually exclusive classes, but that we occupy a place along a "patterned continuum." Weiss (1973) who used the concept of "patterned continuum" has dealt with these issues and their far-reaching consequences for society in a most perceptive and penetrating manner.

Hassenstein (1954, 1971), who instead of 'patterned continuum' used H. Rickert's term 'heterogeneous continuum', also emphasized the fuzziness of so-called biological entities (or classes) which are no entities because they have no boundaries. Since in such cases definitions would lead to arbitrary and artificial delimitations, Hassenstein proposed to refrain from definitions and to use concepts as "injunctions", i.e., as fuzzy, nondefinable concepts. What is meant, can be illustrated diagrammatically (Fig. 4.2).

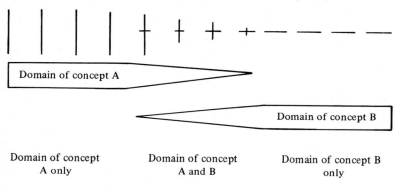

Fig. 4.2. Concepts as injunctions. (Slightly modified after Hassenstein 1954, p. 199)

An injunction is characterized by an area in which the applicability of the concept fades out. Areas of partial applicability of concepts may overlap as is indicated in Fig. 4.2. In that case the area of overlap has to be characterized by two (or more) concepts each of which applies partially. Examples of biological concepts that may serve as injunctions (and hence are not defined) are "individual," "plant" and "animal," "healthy" and "sick," "domestication," "innate" and "learned," "territorial behavior," "species," and "life" (see Hassenstein 1978). Properties may, of course, be enumerated for each of these concepts, but in the fuzzy margin of the concept those properties apply only partially. Hence, neither monotypic nor polytypic definitions are adequate. Fuzzy set theory may perhaps represent such situations in a mathematical fashion.

The notion of types may be used in the sense of sharply delimited, mutually exclusive classes or as injunctions. The latter use appears to be appropriate in many biological instances where sharp boundaries are lacking.


## 4.1.10 Fundamental Conclusions and Consequences

(1) Conceptualization is a prerequisite for science, i.e., science has to be conceptual. (If a nonconceptual science ever should be developed, it would

be quite different from anything that is considered science today. Maslow's (1966) discussion of what he calls "fusion-knowledge" is highly relevant in this respect).

(2) The conceptual approach is not restricted to science, but is also characteristic of philosophy, everyday life and to some extent even religion. Therefore, the conclusions to be drawn regarding the use of concepts apply not only to science, but also to our personal lives. This means that understanding the use of concepts in science can generally educate us.

(3) Concept formation removes us one step from reality through the process of abstraction which means separating certain features from the welter of experience (Langer 1964). "Nothing could be further from the truth than that there is complete correspondence of structure between all experience and all language, or even between any limited aspects of language and experience" (Bridgman 1936, p. 23).

(4) Through abstraction we create things (including events) that do not exist as real entities independently of our conceptualizations. Thus, entities are the result of "selective attention" (Watts 1970, p. 57). For example, our bodies do not really exist as separate entities. "Skin is as much a joiner as a divider ... the bridge whereby the inner organs have contact with air, warmth and light" (Watts 1970, p. 55). "Basically, everything is one" (Barbara McClintock, quoted by Keller 1983, p. 204).

(5) What applies to things, is also true for classes of things: they do not exist independently of our conceptualization.

(6) As a result of the conceptual approach one aspect of reality escapes us, namely the unity or wholeness. In conceptualizing we chop up the unity into digestible portions and thus create a fragmented world.

(7) A fragmented vision and experience of the world and ourselves will eventually lead to physical and mental illness. The prevalent feeling of loneliness and all kinds of neuroses are some of the consequences of fragmentation.

(8) Fundamental questions facing us today especially in Western and westernized cultures are: How can we regain a more balanced life that is whole? How can we be healed? For "to be healed is to be made whole" [Bohm (1980, p. 3) who also pointed out that the word 'health' is derived from 'hale' which means whole]. How can we see reality as it is, undivided by categories and abstractions? There is no single, simple, and foolproof answer to these questions. Meditation in its many ways as sitting and dynamic meditation calms the mind (which normally is jumping from fragment to fragment) and may show the way to the Unnamable (see, e.g., Krishnamurti 1970; Rajneesh 1976).

(9) Although conceptualization leads to fragmentation and may create devastating problems if indulged in one-sidely, it represents an aspect of

reality. Adequate concepts, like good maps, may help to orient and guide us. Awareness of the process of conceptualization leads beyond the latter. Thus, conceptualization in awareness need no longer be destructive.

## 4.2 Classification

### 4.2.1 Classes

Scientific statements are in terms of concepts since science is conceptual. Among the kinds of concepts used, class concepts play a very important role. It is probably no exaggeration to state that most, if not all, scientific statements, whether they are singular or general propositions, contain class concepts. For example, the fact 'this tree is dead' contains the class concept "tree." The laws of genetics are based on the class concept "gene." Many other biological statements ranging from molecular biology to taxonomy and ecology are made with reference to certain taxonomic groups (taxa) such as species, families, phyla, or kingdoms. Thus, in making scientific statements it seems to be unavoidable that we distinguish classes. If the mere use of class concepts is called classification, classification is all-pervasive in biology, be it in systematics, ecology, physiology, genetics, or biochemistry. However, often the notion of classification is used in a more restricted sense, namely in the sense of classifying classes to form systems of classification. Such systems can be constructed with classes of different content. For example, the classes can be classes of biochemical compounds (in the field of comparative biochemistry), classes of behavioral patterns (in ethology), classes of functions (in physiology), classes of structures (in morphology), or classes of individual organisms and groups of organisms (= taxa) (in taxonomy or systematics).

According to Simpson (1961, p. 7), systematics is defined as "the scientific study of the kinds and diversity of organisms and of any and all relationships among them." "Relationship" may be understood in a very general way, including similarities as well as phylogenetic connections. Taxonomy, also according to Simpson (1961, p. 11), is defined as "the theoretical study of classification, including its bases, principles, procedures, and rules." Thus, systematics is concerned with the relationships among organisms and groups of organisms, whereas the subject of taxonomy is classifications. It should be noted, however, that not all biologists accept these distinctions.

As far as classes in general are concerned, they can be conceived as monotypic or polytypic. Thus, species (of plants and animals) usually have been considered as monotypic or polytypic classes. Different kinds of properties, i.e., different criteria, are used to establish species. Accordingly, different species concepts are distinguished (see, e.g., Slobodchikoff 1976; Grant 1981). The major species concepts are the (holo)morphological (phenetic) concept, based on overall similarity, and the biological concept that uses the common gene pool due to reproductive isolation as the main criterion. Both major concepts have been modified and extended in various ways. Thus, as an extension of the biological concept, an evolutionary and an ecological concept have been introduced (Simpson 1961; Van Valen 1976). Still other ways of classifying species concepts have been proposed (e.g., Mayr 1970; Sokal 1974; Grant 1981). A synthesis of different concepts using different criteria also has been attempted (Wagner 1984). Small (1979) discussed many of the different species concepts particularly with regard to marijuana and emphasized the social and legal implications of species delimitation.

The literature on species concepts is enormous. Some of the philosophical issues at stake are the following. For the conceptual realists the question is which of the concepts is the real one. They usually opt for some kind of biological concept (see, e.g., Kitts and Kitts 1979). For those who accept the view of concept formation as presented in the preceding section, the question may arise which of the various concepts is the most adequate one. However, it need not be assumed that one of them must be superior as far as adequacy is concerned; they may in fact be equally adequate representing complementary aspects of the "patterned continuum" of nature. Much evidence exists for this view (see below) that results in a pluralism which means "that a variety of species concepts are necessary to capture the complexity and variation patterns in nature" (Mishler and Donoghue 1982).

An important contribution to the philosophy of the "species" was made by Ghiselin (1974) and Hull (1976, 1980). These authors consider a species not as a class, but as an individual (see also Mayr 1976). Thus, what we normally consider as an individual plant or animal organism, in terms of this concept, is seen as an integrated subunit (part) of one individual species. For example, *Homo sapiens* is an individual according to this view, and what we usually call 'individual human beings' are parts (subunits) of this individual, the human species. An empirical basis for this view of species is the common gene pool that provides spatiotemporal continuity and integration between the organisms of a species [see, e.g., Richmond (1979, p. 247) with regard to genetic integration in bacteria]. Since the organisms also have a relative autonomy, class concepts and the new individualistic concept of the species may be considered complementary to each other in a pluralistic outlook as indicated above. For a critical appraisal of the individualistic species concept see Caplan (1981), Hull (1981b), and Mishler and Donoghue (1982).

If one accepts the individualistic species concept as a new valid concept, which I think it is, then a more general conclusion may be drawn. As the organisms belonging to one species are integrated with each other, so species are continuous with each other as a result of their evolution. Even if one accepts Eldredge and Gould's (1972) "punctuated equilibria," i.e., the notion that the evolution of species is relatively rapid and abrupt, one cannot deny that the transition from one species to another is a relatively continuous process which, in time, extends the gene pool of one species to that of another by a more or less gradual (or abrupt) change of the former. Only an observer who considers simply a cross-section of phylogeny at one particular time, i.e., a time-slice of phylogeny, may come to the conclusion that many species are discontinuous. If phylogeny is seen in its space-time extension, then the continuity between species is obvious. This means, however, that if we accept the individualistic species concept, we may go further to consider as one individual the totality of species that share in a common gene pool or exhibit other ways of genetic integration (see, e.g., Richmond and Smith 1979; Margulis and Sagan 1985). If now, in addition to genetic integration, the interaction and continuity of organisms with the environment is taken into consideration, then we can see the whole world as one integrated system (see, e.g., the Gaia hypothesis proposed by Lovelock 1979). We arrive at a scientific view of unity or oneness that is fundamental in mystical experience.

### 4.2.2 Systems of Classification

One question that arises with regard to systems of classification is: how do we construct them? Due to the biophilosophical studies of many taxonomists much has been learned about taxonomic methodology during the last decades. Particularly the analytical studies of the numerical taxonomists (Sneath and Sokal 1973) should be taken into consideration, regardless of whether one accepts the philosophical and biological tenets of the numerical taxonomists, which have become increasingly sophisticated, diversified and comprehensive; they may even include evolutionary aspects and thus are not necessarily restricted to phenetic classifications based on equal weighting of characters (see below).

The following steps may be distinguished in the construction of a phenetic system of classification: (1) the choice of classes (that will be classified) with their members and the distinction of characters with their states, (2) the determination of the degree of similarity or difference of the classes, and (3) the arrangements of the classes in a system of classification according to the degree of similarity or difference.

To 1. A character has at least two states (or attributes). An example of a two-state character is "leaf surface hairy," or "leaf surface not hairy." Hairiness is one state, nonhairiness is the other state. It is evident that in certain cases a multistate character would be more appropriate, such as, for example, the three-state character: "leaf surface not hairy," "leaf surface slightly hairy," and "leaf surface densely hairy." A continuum of states naturally would be the most adequate representation in at least certain continuous samples.

Characters and their states are the result of conceptualization. This conceptualization provides the foundation for what is considered the same character and the same state in different organisms. Sameness is a fundamental issue. We know that sameness in the sense of identity does not exist. Thus we never step into the same river twice because the second time neither the river nor we are exactly the same as before. "Panta rhei," i.e., all is flow (Heraclit), means that the water of the river changes constantly, and, as modern biology has shown, most of the cells and molecules of our body are replaced at a rapid rate. People who suffer from an "identity crisis" should realize that there is no identity in this world because everything is more or less in flux. What appears to be stable is only so because it changes relatively slowly with respect to faster transformations in the environment. Thus, we must conclude that what is considered to be the same is never exactly the same. It can at best be an approximation to sameness. Hence, the very basis of the classificatory process that assumes sameness of characters is already an oversimplification. It is only through abstraction that sameness emerges. The task, of course, is to abstract in the most adequate way.

Showing the sameness of characters is usually referred to as homologization (according to one definition of homology). I think that homologization of this kind, i.e., the establishing of total correspondences (or 1:1 correspondences) of characters, may be considered adequate in a number of cases. In other cases the correspondence of characters is rather partial. In these cases the traditional concept of homology implying sameness should be replaced by a concept of partial homology which I have termed a semi-quantitative homology concept (Sattler 1966, 1984).

Homology can be considered as the central concept of comparative biology and as the most basic concept in taxonomy because homologization has to be carried out as the first step in the process of classification. It is ironical that the central concept of comparative biology has been treated predominantly as a qualitative concept which implies that a character is *either* homologous with another one *or* not. If, as suggested, homology were conceived as a comparative (= semiquantitative) concept, this would render comparative biology a truly comparative discipline. A further step to be taken would be the quantification of homology (see, e.g., Sneath and Sokal 1973). Fuzzy set theory may provide the appropriate tool in this respect. It would allow us to see a quantitative relation between the so-called same characters and others that are different. Total homology would then turn out to be the borderline case in which the degree of homology (= degree of similarity) approaches 100% without ever reaching it completely since absolute sameness does not exist. (For a critique of phylogenetic or evolutionary homology concepts see Sattler 1984.)

To 2. The second step in the process of classification involves the determination of the degree of similarity or difference between the chosen classes (or individuals). This can be done quantitatively through the use of a similarity (or difference) coefficient. As a result, the classes (or individuals) can be arranged in a matrix according to their degree of similarity or differences (see, e.g., Sneath and Sokal 1973). The practising taxonomist often does all that intuitively.

To 3. The third step is the construction of the system of classification on the basis of the matrix of similarity or difference. An array of quantitative techniques can be used to achieve this clustering (see Sneath and Sokal 1973). Again, this is often done intuitively.

It would be naive to assume that the system of classification obtained in this way represents the System of Nature. As pointed out already, the first step, which involves homologization and the introduction of character states, may lead to considerable oversimplifications and distortions of the natural affinities. The second and the third steps add other problematical decisions. There are several coefficients of similarity and difference. In some cases they may give similar results, in other cases they lead to discrepant conclusions. So, which coefficient are we to use? All we can say in the case of discrepancies is that the similarity or difference obtained is relative to the coefficient used. The same applies to clustering techniques. They may give deviating classifications in at least certain cases. Hence, even the best system of classification reflects not only the order of nature (if nature has any order) but also the methodology of classifying. It would be presumptuous to believe that there is one methodology that is in total harmony with nature so that nature would describe herself. Even if this wishful thinking were justified, how could we decide whether it is the methodology of the taxonomist x, y, or z that reflects nature as she is?

The preceding account of the methodology of classification is not complete. It is actually still more complicated and involves additional assumptions and problems. For example, in the determination of the degree of similarity or difference, equal weight can be given to all characters or characters can be weighted differentially. Equal weighting is characteristic of the neo-Adansonian school of numerical taxonomists (see Sneath and Sokal 1973). It leads to phenetic systems that reflect the overall similarity of the classes. Although differential weighting of characters is also possible in phenetic taxonomy, it is mainly practiced by taxonomists who insist that adequate systems of classification must be either phylogenetic or at least consistent with phylogeny (see, e.g., Simpson 1961; Mayr 1969, 1981;

Eldredge and Cracraft 1980;Platnick and Funk 1983). The difficult question that arises for these taxonomists is how to determine the proper weighting of characters. Much disagreement and vagueness exists in this respect and as a consequence phylogenetic systems of classification are often rather speculative, especially in taxa for which fossil evidence is lacking or poor. Thus it is not surprising that even the most enthusiastic proponents of phylogenetic systems, such as, for example, Simpson (1961) could write the following: "Taxonomy is a science, but its application to classification involves a great deal of human contrivance and ingenuity, in short, of art. In this art, there is a leeway for personal taste, even foibles, but there are also canons that help to make some classification better, more meaningful, more useful than others" (quoted after Mohr 1977, p. 125). Some critics of phylogenetic and evolutionary systematics would be even more negative. On the other hand, cladists who have developed a special methodology of phylogenetic systematics are more optimistic as far as phylogenetic reconstruction and classification are concerned (see, e.g., Cracraft 1983).

Pages and pages, articles and articles have been written to show either that phenetic systems are the natural ones or, more frequently, the contrary, i.e., that only phylogenetic systems can be the systems of nature. From what has been said before about the relation of concepts and systems of classifica-

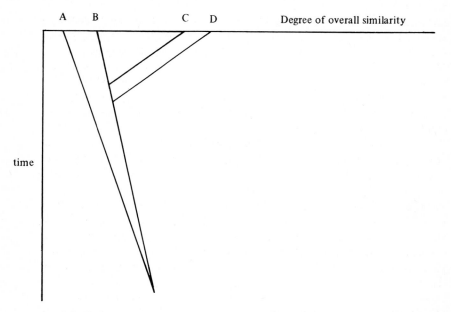

**Fig. 4.3.** Scheme showing a case of discrepancy between phenetic and phylogenetic classification. Explanation in the text

tion to reality, it follows that neither a phylogenetic, nor a phenetic system can be natural in the sense of representing nature as she exists independently of our conceptualization. However, both kinds of classificatory systems present aspects of nature. To some extent these aspects are complementary to each other, to some extent they may even coincide depending on the situation. Figure 4.3 shows a case in which the phylogenetic and the phenetic systems complement each other because they show different aspects of order. According to the degree of overall similarity, which is expressed on the horizontal axis, the classes A and B form one group and the classes C and D another one. However, according to recency of ancestry, i.e., the phylogenetic criterion of divergence in time, the class A forms one group which is distinct from another group consisting of classes B, C, and D. In the latter group, one can distinguish one subgroup consisting of B and C, and another one represented by D. Obviously the two systems are quite different. To claim that only the phylogenetic one represents nature and to declare the phenetic one as meaningless (see Mohr 1977, p. 125) does not seem valid. Without a system that expresses the degree of overall similarity, one could neither express the phenomenon of parallel evolution which is the case in the evolution of classes C and D (Fig. 4.3), nor could one assess the phenomenon of divergent evolution at different rates which is shown in Fig. 4.3 where A and B diverge much more slowly from each other than C and D from B. In any case Fig. 4.3 shows that the overall degree of similarity need not coincide with the degree of recency of ancestry, i.e., the phylogenetic relationship. Hence the two kinds of systems present different aspects of nature and in that sense are complementary to each other. Many cases of such discrepancies have been reported in the research of taxonomists (see, e.g., Simpson 1961; Davis and Heywood 1963; Mayr 1969).

Instead of contrasting phenetic with phylogenetic methodologies, one might distinguish a larger number of approaches to classification. Thus, Vickery (1984), in an analysis of nine populations that represent all the six species of the section *Erythranthe* of the genus *Mimulus*, used the following five approaches: (1) numerical taxonomy, (2) chemotaxonomy, (3) allozyme analysis, (4) DNA/DNA hybridization, and (5) standard cytogenetic approaches (i.e., determination of $F_1$ fertility). All of them produced different classifications of the nine populations (see Fig. 4.4). Vickery (1984) emphasized that each of these classifications can be seen as a different perspective. "Each approach adds to our information about the biology of the group – often permitting distinctions to be drawn between entities that other approaches did not resolve. Clearly, they all are complementary" (Vickery 1984, p. 12). Vickery superimposed all five classifications in one scheme (Fig. 4.5) which, of course, does not yield one synthetic classification. A synthesis does not seem in reach. We have to content ourselves with more or less contradictory perspectives that show different aspects of reality.

94

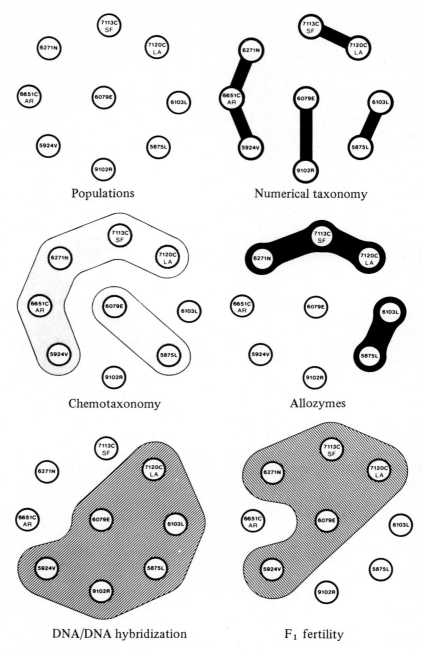

Populations

Numerical taxonomy

Chemotaxonomy

Allozymes

DNA/DNA hybridization

$F_1$ fertility

**Fig. 4.4.** Legend see opposite page

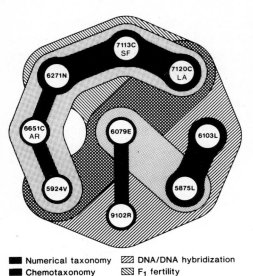

**Fig. 4.5.** Superimpositions of the five different classifications of Fig. 4.4. Apart from the coincidence of a group of two populations according to numerical taxonomy and allozymes (see Fig. 4.4), there is no correlation of groupings. The five classifications represent complementary perspectives that contradict each other more or less. (After Vickery 1984)

■ Numerical taxonomy    ▨ DNA/DNA hybridization
■ Chemotaxonomy        ▧ F₁ fertility
▨ Allozymes

A system of classification may be considered as a theory that is testable by the predictions that can be deduced from it. It seems that good phenetic systems might yield better predictions and confirmations than phylogenetic systems. Reasons for this were given by Gilmour (1940, 1961). He made a distinction between a general purpose system, which is based on many criteria, and a special purpose system, which is based on only one criterion. According to this view, the phenetic system is considered a general purpose system because it is based on many criteria such as morphology, chemistry, physiology, ethology, etc., whereas the phylogenetic system is a special purpose system based solely on the genealogical criterion. On this basis one might argue that a phenetic system has more information content than the phylogenetic system and therefore provides more and better predictions and confirmations. It should, however, be kept in mind that, although the phylogenetic system is based on only one criterion, indirectly it may reflect also

---

**Fig. 4.4.** *Upper left:* Nine populations representing all six species of section *Erythrante* of the genus *Mimulus. Upper right:* classification consisting of four groups (indicated by *black bars*) according to numerical taxonomy. *Center left:* two groups of most similar populations based on chemotaxonomy. *Center right:* two groups of most similar populations on the basis of their allozymes. *Lower left:* Six populations forming one group according to DNA/DNA hybridization. *Lower right*: Five fully inter-fertile populations can be distinguished as a group on the basis of the fertilities of the interpopulation $F_1$ hybrids. (After Vickery 1984)

other criteria. Therefore, it is questionable whether the phylogenetic system contains less information than the phenetic system. The difficulty of making predictions on the basis of a phylogenetic system may be related to other factors (see, e.g., Hull 1974). It is also a controversial issue whether evolutionary theory can make predictions (see Chap. 8).

Systems of classification, which can be conceived as theories, are themselves dependent on other theories including the theoretical assumptions of the methodology by means of which they are constructed. Thus, phylogenetic systems depend on evolutionary theory. One reason why the pheneticists originally objected to phylogenetic systems was their dependence on evolutionary theory. They hoped, at first, to make phenetic systems totally repeatable and objective (real), a dream that did not last long because it was shown that phenetic systems depend on assumptions or theories too, although not on evolutionary theory (see, e.g., Hull 1974).

With regard to the structure of systems, phylogeneticists, and to some extent also pheneticists, tend toward hierarchical systems of classification because hierarchical structure is often perceived as natural. Riedl (1980, p. 108), for example, claims that the natural order of life is a hierarchical one and that our thinking in terms of hierarchies has been selected as an adaptation to the nature of things. This, according to Riedl, explains why hierarchical thinking dominates systems of classification from everyday life and aspects of our society to scientific systems of classification. Although Riedl's argumentation may appear convincing at first sight, one may pose a number of questions. Is it possible that the hierarchical order is projected into nature by us and that therefore we find hierarchies everywhere as Riedl contends? Is it possible that due to our cultural background, which has been greatly influenced by hierarchical thinkers, we have become conditioned to think in terms of hierarchies? Riedl (1980) reports how quickly we learn if there is sufficient reinforcement. Now we know that especially in our Western culture the hierarchical reinforcement is constantly at work in almost all aspects of our lives. It would be indeed surprising if we would not think hierarchically. Therefore, the fact that our lives and our thinking are permeated by hierarchies does not at all mean that nature independently of us has to be hierarchical.

In order to test the adequacy of the hierarchical notion, we should observe nature in as unbiased a way as possible. Such an observation reveals (1) that there are cases where hierarchical thinking leads to difficulties and inconsistencies, and (2) that such cases are more adequately modeled by nonhierarchical systems such as nets. Nets are known not only in systematics (see, e.g., Sneath and Sokal 1973, p. 260; Stace 1980, p. 155), but also in morphology (see Chap. 5 on plant morphology), in neurobiology (neural networks), in morphogenesis (morphogenetic networks), ecology (e.g., food webs), etc. Probably, one could find examples in all biological disciplines. Furthermore, in many other domains of human activity the terms net and network have been applied successfully. Cybernetics and systems thinking utilize the concepts of feedback loops and networks. Nets are not only known in the physical sphere, but also in the mental realm. Novak (1971, p. 55) characterized the self as a net and Hesse (1974) developed a network theory of scientific methodology. The scheme of Laszlo, which I have modified slightly (see Chap. 1, Fig. 1.7), is also a network model of the scientific enterprise. A brief scan of bibliographic aids such as *Biological Abstracts* reveals that nets and networks have been reported in many instances and in many different disciplines (see Chap. 6 for a discussion of network thinking).

I think that nets are often a more adequate, although more complex, representation of nature than hierarchies. In these cases, hierarchies can be considered to be a simplified and an impoverished version of nets in which some relations or interactions have been ignored. These relations have been indicated by dotted lines in Fig. 4.6. Note that a net results if the dotted lines are drawn out, i.e., are added to the others.

The reason why the notion of hierarchy works to a certain extent may be that in some cases the dotted lines represent weaker relations or interactions so that they may be negligible without a great loss of information. Thus hierarchies may be considered to be useful approximations in a number of cases, although nets seem to be the more adequate model in many cases. It is, of course, possible in certain cases such as phylogenetic trees to abstract in such a way that trees and hierarchies are adequate representations.

The challenge of modern systematics is to adapt our methodology of classification to nature so that the resulting systems will become more adequate. We distort nature if we impose rigid class concepts where she is fuzzy

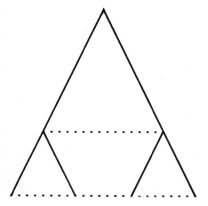

Fig. 4.6. Diagram showing the relation of a hierarchy and a net. The elimination of the *dotted lines* transforms a net into a hierarchy (for an example from plant morphology see Chap. 5, Fig. 5.4)

and continuous. In those cases injunctions and fuzzy set theory may be more appropriate. Classification thus turns into ordering. Ordering is a broader concept than classifying by means of monotypic or polytypic classes in hierarchical or nonhierarchical systems. Ordering can lead to yet other systems. For example, clinal variation, reported in the biological literature, in a simplified and formalized representation may have the following order:

$$abc$$
$$bcd$$
$$cde$$
$$def$$

If the four elements (abc, bcd, etc.) represent individuals, this clinal system cannot be said to consist of classes, yet it exhibits order. A well-known example is the change of vegetation along a gradient(s), such as increasingly restricted precipitation (see Whittaker 1973; Tivy 1982, p. 197). Since the change in vegetation structure is continuous, the delimitation of vegetation types is arbitrary. In such cases ordination provides a more natural representation of the phenomenon than classification. Ordination is defined as "the arrangement of samples in relation to environmental gradients as a basis for gradient analysis" (Whittaker 1973, p. 6). A great variety of ordination techniques have been developed (see, e.g., Whittaker and Gauch 1978).

Although in the preceding discussion mainly classes of individual organisms and groups of organisms were considered, all sorts of other entities can be classified, such as, for example, processes, functions, chemical compounds, hypotheses, systems of classifications, methods of classification, ideas, philosophies, ideologies, religions, life-styles. Again, it may often be more appropriate to look for order that is not expressed by a system of classes but rather by a net of fuzzy sets or injunctions. Even the latter may create distortions (see Chap. 2 on the chaotic aspect of nature), but they may be more adequate than conventional classes and hierarchical systems. The latter will, of course, retain a certain practical usefulness, especially if we are aware of their limitations.

It should not be overlooked that the issues of classification and ordering may have great social relevancy inasmuch as concepts such as "IQ," "race," "ideology," or "religion" are concerned [for a case study on marijuana showing the social and legal implications of the species problem see Small (1979)].

## 4.3 Summary

*Concepts.* Science, like philosophy, is conceptual. Hence, an understanding of the meaning and formation of concepts is a prerequisite for the understanding of scientific statements. Concepts are abstractions. Abstractions constitute a selection of feature(s) from direct experience. Since there are many ways of selecting feature(s) from the richness of experience, many different abstractions, i.e., concepts, can be formed from the same background experience. A much debated question is whether at least some of these concepts are real, i.e., represent reality as it is, or whether all of them are arbitrary. Instead of this alternative, it is suggested to ask to what extent a concept corresponds with reality. The closer the correspondence, the more adequate is the concept. Complementary concepts refer to different aspects of reality. One of them need not be more adequate than the other one. Application of concepts beyond their domain of applicability leads to meaningless questions and statements. Different classifications of concepts exist. For example, qualitative, comparative (= semiquantitative), and quantitative concepts may be distinguished, or, according to another classification, individual concepts, class concepts, relation concepts, and quantitative concepts exist. The distinction between a term and a concept is important because one term may designate more than one concept or vice versa. 'Define your terms' is an important maxim, yet one should realize that all terms are eventually defined by primitive terms that can not be defined, but are understood intuitively. Thus, the whole conceptual edifice of science basically rests on intuition. Operationism is a philosophy that maintains that concepts are defined by operations that determine their application. This insistence on operational definitions guarantees that concepts are not applied beyond their domain of applicability. The disadvantage is a lack of generality and a failure to recognize certain phenomena. Although there are no absolutely biological kinds of concepts, historical, functional, and polytypic (= polythetic) concepts are typical for biology and of less importance in physical sciences, such as physics and chemistry. The great significance of polytypic concepts (classes) is underlined. Fuzzy set theory and injunctions can be seen as a further extension and revolution beyond the notion of polytypic concepts.

In a general conclusion it is pointed out that concept formation, which is so fundamental to science, philosophy, everyday life and to some extent even religion, removes us one step from reality. Through the process of selective attention (abstraction), things and classes of things (including events) are created. Consequently, the unity of the universe is fragmented into digestible bits. This fragmentation is useful inasmuch as it is in partial correspondence with reality and thus provides conceptual guidelines for our orientation in the world. However, the vision and experience of unity is also necessary for a fulfilled, rich personal life and the harmonious survival of humankind. Meditation may be one means to calm the troubled mind dispersed in a fragmented world of concepts and to regain awareness of the wholeness of life.

100

*Classification.* In the widest sense of the term, we are classifying whenever we use class concepts. Since most, if not all, biological statements contain class concepts, classification in this sense is an all-pervasive activity. The classes may be monotypic or polytypic. As an example of class concepts different species concepts are mentioned. The individualistic species concept supersedes the notion that a species is a class of individual organisms. According to this concept, all members of a species are integrated parts of one individual due to a common gene pool. Since species are continuous during evolution and since living systems are integrated with the so-called abiotic environment, unity is all-pervasive. In a sense this conclusion of modern biology parallels the experience of oneness by mystics.

In a narrower sense, classification refers to the construction of a system of classification. Three steps may be distinguished in the construction of a phenetic system: (1) the choice of classes (that will be classified) with their members and the distinction of characters with their states, (2) the determination of the degree of similarity or difference of the classes, and (3) the arrangement of the classes in a system of classification according to the similarities or differences. With regard to the first step, establishing equivalent characters implies the postulate of sameness (conventional homology). Since absolute sameness does not exist, a replacement of "sameness" by "similarity" would be appropriate. This means that the conventional qualitative homology concept should be replaced by a comparative (semiquantitative) or quantitative homology concept. With regard to the second and third step in the construction of a system, various other difficulties arise. Different techniques such as different similarity (or difference) coefficients may at least in some cases produce different systems of classification. The weighting of characters also influences the resulting system. Neither phenetic nor phylogenetic systems can be considered natural in the sense that they present the system of nature (if there is one). Both kinds of systems are complementary, representing different aspects of nature. Both are theory-dependent and thus constantly open to doubt and change. Although the form of most systems is hierarchical, it need not be so. Nets often appear to be more adequate representations of nature, although hierarchies have a limited usefulness. As we adapt our methodology more and more to nature, ordering may have to supersede classifying.

Both classifying and ordering apply to many entities in biology as well as in everyday life. Thus we may classify and order, for example, structures, chemical compounds, processes, groups of organisms, systems and methods of classification, hypotheses, ideas, philosophies, ideologies, religions, and life-styles. The social relevance of classification and ordering is especially obvious with regard to issues such as I.Q., race, ideology, and religion.

# 5 Comparative Plant Morphology: A Biophilosophical Case Study

> "We shall confine ourselves here to one chosen approach
> – that of morphology. This may seem a narrow road, but,
> rightly conceived, it should, like other biological paths,
> lead us towards infinite issues" (Arber 1950, p. 1)

The purpose of this chapter on plant morphology, a discipline in which I have been carrying out practical and theoretical research for over 20 years, is at least twofold: (1) to illustrate the major concepts and ideas of the preceding chapters in terms of one particular biological discipline, and (2) to explicate the philosophical foundations of one biological discipline with which I am familiar. I hope that through this interaction between a specific biological discipline and biophilosophy both will profit. Plant morphology may become more transparent through a clarification of its philosophical foundation which may raise into conscious and explicit form assumptions of which we may not be fully aware. Biophilosophy will be tested with regard to its practical relevancy.

The presentation of this case study follows that of the preceding chapters. Only in a few instances have I departed from that organization. For example, the various kinds of hypotheses are not illustrated at the beginning of Chap. 1, but in a later context.

## To Chapter 1: Theories and Hypotheses

The aim of plant morphology is to gain scientific knowledge of plant form and structure by means of singular and general propositions. As in other branches of science, the greatest degree of generalization is sought, although description at the factual level and less comprehensive generalizations play an important role. One of the most comprehensive generalizations of the comparative morphology of seed plants may be formulated as follows:

Seed plants are composed of three mutually exclusive kinds of organs: roots, stems and stem homologs (= caulomes), and leaves and their homologs (= phyllomes). Branching of caulomes occurs in the axils of the phyllomes [Sattler (1974a); for a diagrammatic illustration see Fig. 5.1].

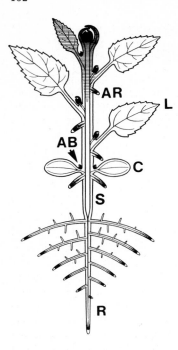

**Fig. 5.1.** Diagrammatic representation of a seed plant indicating the three kinds of organs, root (*R*), stem (*S*), and leaf (*L*). Adventitious roots (*AR*) are born on the stem. Axillary buds (*AB*) are present in the axils of the leaves. The first formed leaves are the seed leaves (cotyledons, *C*) (After Troll 1954)

This generalization reduces the enormous diversity of form and structure of seed plants to three kinds of elementary units (i.e., three kinds of organs). Flowers are also interpreted in terms of these units as systems consisting of a caulome (the floral axis) and phyllomes (the floral appendages such as sepals, petals, stamens, and carpels) (see Eyde 1975; Weberling 1981). This generalization has been referred to as the classical theory of plant morphology. It has been accepted by the majority of plant morphologists (e.g., Troll 1937; Eames 1961; Weberling 1981). Needless to say, it has been interpreted in different ways by different authors (see Cusset 1982) and a growing number of botanists accept it only to a certain extent or as a "workable rule of thumb" (Tomlinson 1982, p. 165). Is it a theory, or should it be called a concept, a law, a rule, a model, or even a paradigm? If it is considered a concept, it is a rather complex concept that comprises a whole set of concepts. Thus, it is a conceptual framework that forms the basis for much work in plant morphology and other disciplines based on it. The concept of "model," which is more and more used in a great variety of meanings, can be easily applied to this generalization (see, e.g., Bertalanffy 1965, 1975). Since the latter is well confirmed and so comprehensive that it comprises laws or rules (see below), it might also be termed a theory. And inasmuch as it is the foundation of much work in the comparative morphology of plants, it might even be considered a paradigm (see, e.g., Mohr 1977).

Biological generalizations are not always easily categorized in terms of the various kinds of generalizations distinguished in Chap. 1. Evidently, this is the case in plant morphology with regard to the above generalization. In this chapter, I shall refer to it as the classical model keeping in mind that another categorization might be equally appropriate. It is clear, however, that it is not a fact (i.e., a singular proposition), since it is a highly general statement referring to seed plants (or even the majority of vascular plants).

How do we gain scientific knowledge in plant morphology? In the case of the classical model, how was it discovered or invented? One might think that such a generalization could be arrived at inductively. Thus at first some plants consisting of the three kinds of organs and showing axillary branching would have been observed. These facts then would have been generalized inductively and in this fashion the classical model would have been generated. However, general analytical considerations as well as the actual historical events leading to the development of the classical model contradict this assumption of inductive generalization (see, e.g., Troll 1926; Arber 1946; Cusset 1982). Although it is possible that a less general proposition, such as the statement that all trees have leaves, might be formed inductively, the invention of a generalization with the scope of the classical model required intuition and imagination. Actually, during his Italian trip (1786-1788), it flashed upon Goethe's inner eye that all appendages of plants are homologous, i.e., seed leaves, scale leaves, foliage leaves, bracts, sepals, petals, stamens and carples are all essentially the same (Goethe 1790; Arber 1946). Observations, such as that of the now famous palm (*Chamaerops humilis* L.) in the Botanical Garden at Padua in Italy, were important. But none of these observations alone inductively generated the classical model. Goethe's intuition was necessary to arrive at this generalization. If, however, induction is understood in a very broad sense so that it is part of the intuitive process, induction played a role. Once the idea of the classical model had been conceived, it could be tested and confirmed according to the hypothetico-deductive model (see Figs. 1.3 and 1.6, Chap. 1). Since it has been confirmed over and over again up to the present time, it is considered a highly confirmed model or theory. Nonetheless, it is not proven. New observations that contradict the model may be made any day. Actually, many contradictory observation are already known (see, e.g., Sattler 1974a). Thus, intermediate structures that do not satisfy the postulate of three mutually exclusive kinds of organs have been reported (see also Cusset 1982, pp. 42-46). Furthermore, deviations from axillary branching have become known (for examples see Sattler 1974, 1975; Dickinson 1978; see also Figs. 5.2, 5.3). Do such contradictory observations falsify the classical model? They do not if falsification is under-

**Fig. 5.2.** Portion of a twig of *Dulongia acuminata* H.B.K. (= *Phyllonoma ruscifolia* Willd.) showing inflorescences born near the tips of leaves (Redrawn by Mrs. Eva Krivanek after the original by P.J.F. Turpin in A. von Humboldt, A. Bonplaud, and K.S. Knuth: Nova genera et species plantarum, 7, plate 623. Paris, 1825). (Reproduced by permission of E. Schweizerbart'sche Verlagsbuchhandlung)

**Fig. 5.3.** Drawing of a longitudinal section of a shoot tip of *Phyllonoma integerrima* (Turcz) Loes. showing the shoot apex (*A*) and two leaf primordia (*L*). The older leaf primordium *to the left* shows the inception of the inflorescence (*arrow*) in a position far removed from the axil (*arrow-head*). [Drawn after a microphotograph by Dickinson and Sattler (1974, Plate 2B)]

stood in an absolute sense. The contradictory observations, although factual, cannot be taken as absolute. Facts also contain a hypothetical element. For example, upon closer inspection an intermediate structure might turn out to be a typical representation of one of the three kinds of organs. Even if we assume that the contradictory observations are correct, falsification would be unattainable because we have to rely on auxiliary hypotheses. For example, vegetative or reproductive branches inserted on leaves contradict the above formulation of the classical model only if we accept the auxiliary hypothesis that these branches are initiated in this position. In a number of cases it was shown, however, that this auxiliary hypothesis is not valid, i.e., the branches are formed in the leaf axil and then are secondarily displaced onto the leaf during the development of the latter (see, e.g., Dickinson and Sattler 1975). However, in other cases, this auxiliary hypothesis has been confirmed, i.e., the branches are initiated on the leaf in a position that is contradictory to the classical model (see, e.g., Dickinson and Sattler 1974 and Figs. 5.2, 5.3). One might be tempted now to conclude that such observations falsify the classical model. However, we cannot be sure that all other relevant auxiliary hypotheses are satisfied. Furthermore, as stated before, we cannot rely 100% on our observations. We have to acknowledge that our facts might be mistaken.

Therefore, it seems to me that all we can conclude is that the classical model has been neither proven nor disproven. It has been confirmed by numerous facts, yet contradictory facts remain. In order to resolve these contradictions I proposed a new model of the shoot (Sattler 1971, 1974) which is based on the old idea of a continuum between morphological categories (see, e.g., Bünning 1977; Cusset 1982). Specifically, the new model comprises two postulates (Fig. 5.4a):

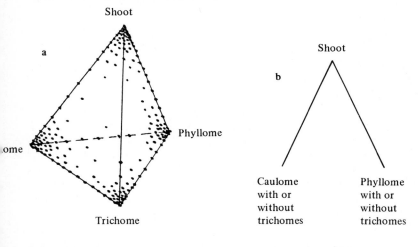

Shoot

a

Phyllome

.ome

Trichome

The position of organs may vary; certain positions such as axillary branching are, however, much more frequent than others

Shoot

b

Caulome with or without trichomes

Phyllome with or without trichomes

Branching is axillary

**Fig. 5.4a,b.** The new shoot model *(to the left)* represents a continuum approach to morphology, whereas the traditional scheme *(to the right)* is rooted in hierarchical thinking and the idea of mutually exclusive categories. According to the continuum model (a) the four typical structures that occupy the four corners of the pyramid are much more frequent than intermediates. The frequency is indicated by the density of dots

(1) there is a continuum between the following four categories: shoot (including the reproductive shoot), stem (in the broad sense of "caulome"), leaf (in the broad sense of "phyllome"), and trichome (= hair)
(2) the position of organs is variable.

This is in contrast to the classical model according to which morphological categories are mutually exclusive at any level of the hierarchy and the position of organs is limited to prescribed sites of the plant (Fig. 5.4b). As I pointed out, the proposed model "can be looked at mainly as a relaxation of the classical model, namely the removal of the mutual exclusivity of classical categories and the restricted positional relations" (Sattler 1974, p. 378). In this sense, the proposed model can also be seen as a modification or extension of the classical model. It is significant that besides morphologists of the last century and the first half of our century, a number of present day botanists have approached plant morphology in a way similar to and/or compatible with my proposals (see, e.g., Lorch 1958; Croizat 1960, 1962; Meyen 1973, 1978, 1984; Jong and Burtt 1975; Jong 1978; Cusset 1982, Sect. 5; Rutishauser 1981, 1983, 1984).

The continuum model appears to be in agreement with all facts that support the classical model. In addition, it is corroborated by those facts that contradict the classical model. One should assume, then, that the scientific community would opt for this model. However, this is not the case. Although an increasing number of my colleagues prefer the continuum model, many are undecided or opposed; thus the botanical community continues to embrace at least to some extent the classical model. To understand this situation we have to consider the interplay of the three kinds of factors in Laszlo's systems model of scientific methodology (Fig. 1.7, Chap. 1). The first kind of factor constitutes the empirical input. Since to some extent the empirical input of plant morphology contradicts the classical model but agrees with the continuum model, one should expect the replacement of the classical model by the continuum model. This is not the case because of the second and third kind of factors. The second kind of factor is the ideals of science which Laszlo represented by two basic values. In the present case these values are in conflict. According to the value of empirical adequacy, the continuum model should be favored. However, in terms of the value of integrative generality many morphologists tend toward the classical model. They feel that the classical model is simpler, more elegant and neater than the continuum model. The reduction of the whole diversity of seed plant form to only three kinds of organs has strong appeal. I suspect that the number three represents a special value. Trinities are venerated in science as well as in some religions.

In conjunction with the ideal of integrative generality the third kind of factor represented by r/g also works against the replacement of the classical model. The resistance factor (r) to change prevents the recognition of anomalies, i.e., contradictory facts, in three ways: (1) Many anomalous data are not perceived. For example, intermediate structures are often seen as extremely modified organs. Thus the phylloclade of *Ruscus aculeatus*, which is intermediate between a leaf and a branch (Sattler 1984, p. 388), usually is only perceived as a highly modified branch (e.g., Kaussman 1955). (For other examples see below). (2) Other anomalies are sometimes interpreted by ad hoc hypotheses. For example, one way of fitting the epiphyllous inflorescences of *Phyllonoma integerrima* into the axillary position as prescribed by the classical model is by means of the ad hoc hypothesis of congenital fusion (see Dickinson and Sattler 1974). According to this hypothesis, the epiphyllous inflorescence is inserted in the leaf axil and is congenitally fused to the leaf. The failure to observe this fusion does not contradict the hypothesis because, in this instance, congenital fusion is a kind of fusion that cannot be observed (Sattler 1978b). This means that facts cannot contradict the interpretation by congenital fusion. In other words: congenital fusion is immune to empirical evidence. Consequently, the classical model is immune to contradictory organ positions if the ad hoc hypothesis of congenital fusion is invoked in a fashion as in the case of *Phyllonoma integerrima*. A model or theory that is immune to facts ceases to be scientific because it is no longer testable. Now I do not claim that the classical model is generally unscientific. But inasmuch as it is protected by ad hoc hypotheses it becomes unscientific. Unfortunately, ad hoc hypotheses are not easily detected and they are not sharply delimited from acceptable hypotheses. (3) Finally, anomalies that contradict the classical model are often suppressed. They are not mentioned in most textbooks. Thus the student is given the deceptive impression that the classical model is free of contradictions. As an excuse for the suppression of anomalies it has been said that they would confuse the student. Is it better to be misled than confused? I am, however, not at all convinced that confusion would be inevitable. Even in a text for beginners one could easily add that the classical model is not free of contradictions. This need not lead to confusion, but to a more realistic perception of the present situation.

Anyone who knows the situation in plant morphology can see that the resistance factor operates strongly against the replacement of the classical model. As indicated in Laszlo's model, it can be counterbalanced by the "genius of innovation" (g). Resistance can be overcome with the proposal of a new model that is attractive to the scientific community. To be attractive, the new model has to satisfy the value of integrative generality, i.e., it has to be simple, elegant, and neat. The problem that the continuum model

108

faces is that many botanists consider it to be neither simple, nor elegant, nor neat. The challenge then is to propose either a different formulation of the continuum model or a different model that will be attractive enough to counteract the resistance to change. Additional well-documented studies of contradictory cases might also help to promote change in theoretical plant morphology.

The situation concerning the impact of values such as simplicity actually is even more complex than so far presented. What to one morphologist may lack simplicity to another may appear simple. For example, the idea of a continuum may be considered simple by one person, yet complex by another. If a morphologist who finds the continuum model simple and attractive would incorporate the continuum model into a widely read textbook, this might constitute an important step toward its acceptance. Other textbook authors, who often copy from existing textbooks, might then include the continuum model into their books and thus propagate it. Similar propagation may occur with regard to reference books and research papers.

As is evident from my modification of Laszlo's model (Fig. 1.7), acceptance of the classical model may affect at least some of the values as well as the empirical input. For example, it may reinforce the belief that nature is simple and neat in terms of categories. Consequently, facts will be described in this manner and nature may appear less continuous than in terms of a continuum model. The important question is: to what extent does the model influence the description of the facts? There is no total dependence of facts on the model, for even an adherent to the classical model may perceive continuities. But I think that from the point of view of the continuum model one will perceive continuities that are not evident from the classical point of view. As examples, I refer again to intermediate structures. From the point of view of the continuum model they are perceived as such and taken as support for the continuum. In classical terms they might be considered as modification of one or the other kind of organ. For example, the stamens of many plants, upon closer inspection, may turn out to be intermediates between phyllomes and caulomes (Rutishauser 1981, p. 36). Classical morphologists perceive them only as modified phyllomes. Occasionally they are interpreted as modified caulomes (see Rutishauser 1981, p. 34). Another example of intermediates that has been well documented are the phylloclades of *Ruscus* and related plants (Sattler 1984, p. 388; Cooney-Sovetts and Sattler, in preparation). These structures are initiated like axillary branches (caulomes) and then develop like leaves (phyllomes). Hence, in their development they combine characteristic properties of caulomes and phyllomes. A recognition of this phenomenon is significant not only for comparative morphology, but also for principles of developmental and evolutionary biology. Evidently, novelty in development and evolution may arise through combination of features or homoeosis (Lodkina 1983; Meyen 1984). In terms of the classical model it arises only through modification.

Just as the classical model reinforces the values of simplicity and neatness, so the continuum model supports the notion that nature is complex and fuzzy (see also Quine 1964). Simplicity coupled with neatness presents one view of the world, whereas complexity and fuzziness allow for a rather different world view. The continuum model dissolves the neatness of three mutually exclusive kinds of organs and thus points to the fuzziness of nature. Each kind of organ is not sharply delimited toward the others but merges gradually with them. From the classical point of view this continuum is not a simple idea. However, as pointed out already, the notion of simplicity is relative. One might say that in another sense a continuum is simpler than or just as simple as the three mutually exclusive categories. The same relativity exists with regard to elegance. Hence, the notions of simplicity and elegance may also apply to the continuum model.

Although Laszlo's model represents much of the complexity of the scientific enterprise, it neglects a few aspects. One of those aspects is partial theory replacement. In that case the new theory does not completely replace the existing one, but both coexist at least for some time. Thus, in plant morphology, the majority of investigators still adhere to the classical model, but some morphologists have accepted the continuum model. Another aspect neglected in Laszlo's model is that a particular scientist may accept a model or theory in part. For example, with regard to the continuum model the positional change of organ inception (such as branch formation on leaves) has been much more easily accepted by morphologists than the continuum between organ categories. Hence, the change from one model, theory, or paradigm to another need not be as sudden and abrupt as is often envisaged. This brings me to a discussion of Kuhn's ideas according to which normal and revolutionary science are distinguished. If the classical model is considered to be a paradigm or disciplinary matrix, research carried out within this paradigm is normal science (see Mohr 1977, p. 141). For example, the elucidation of the development of a thorn may be normal science. A thorn occupies the position of a branch in a leaf axil. Yet it is puzzling because it looks quite unlike a branch. Developmental studies show, however, that it develops like a branch and becomes a thorn through divergence in later developmental stages. Thus, for the classical morphologist the puzzle is solved within the classical disciplinary matrix [see, e.g., Bieniek and Millington (1967) on *Ulex* thorns].

Replacement of the classical paradigm by the continuum paradigm, if that should ever occur, would amount to a revolution (revolutionary science) because it involves a paradigm change (see Mohr 1977, p. 141; Bünning 1977, p. 7). Would such a paradigm change constitute progress in the sense that the new paradigm would be closer to the truth than the existing one? If the paradigms are incommensurable they are not comparable and hence the question is meaningless. In that case the classical paradigm would focus on one aspect of nature, namely the discontinuous one, whereas the continuum model (paradigm) would emphasize the continuous aspect of nature. Hence, the two views could be considered to be complementary. It would be progress to interpret plant form and structure from both viewpoints. Such progress would be particularly beneficial with regard to tolerance between adherents to different models or paradigms and a generally tolerant attitude in our society.

The incommensurability thesis is, however, debatable and I am not completely convinced whether it applies to the situation in plant morphology. Possibly the continuum model includes the classical model as a special case (see Sattler 1974), since in the continuum model continuities as well as discontinuities can be presented, whereas the classical model does not allow for continuities between organ categories.

According to the dialectical scheme of thesis, antithesis, and synthesis, the classical model represents the thesis, and an extreme formulation of the continuum view that focuses only on the continuum represents the antithesis; the synthesis is the view that plants form a "patterned continuum" (Weiss 1973): this synthetic continuum view recognizes both the continuum and the pattern that could be described as relative articulation of the continuum or as relative discontinuity. In this book as well as in previous publications I have used the terms 'continuum,' 'continuum model,' 'continuum view,' etc., in the synthetic sense. Accordingly, I recognize that each plant shows relative articulations or discontinuities and the distribution of typical and atypical forms is also heterogeneous, i.e., the typical forms are much more frequent than the atypical ones.

Finally, I want to emphasize that the continuum view need not be interpreted as a new paradigm and disciplinary matrix. As I pointed out already, it may just as well be seen as a modification and extension of the classical approach. The classical approach is in fact so plastic that it intergrades with the continuum view. Guédès (1979, p. 135), for example, recognizes positional variability of organs (such as epiphyllous leaves in *Begonia hispida* var. *cucullifera*), while retaining the mutual exclusivity of the three kinds of organs. In other words: he tasks a viewpoint that is intermediate between the typical classical model and the continuum approach. Other morphologists could be quoted whose views combine elements of both the classical and continuum approaches. This means that the two approaches are not sharply delimited from each other, but form a continuum. Any opposition of the two approaches as two mutually exclusive schools of thought is therefore artificial. In addition, opposition and polarization may be detrimental to the discipline of plant morphology and the personal relations between morphologists because it focuses on exaggerated discontinuities while neglecting the common ground and the links that unite.

The preceding discussion was based on the commonly held idea according to which a theory is "a system of logically related propositions" (Bunge 1980, p. 225) that makes empirical claims about the world or some aspect of it. Thus, the classical theory (or model) of seed plants postulates that the whole diversity of these plants is reducible to three kinds of organs in specific positions. If we now formulate the classical theory in terms of the semantic view of theories, it becomes the following definition: "The model of seed plants consists of three kinds or organs in specific relative positions." This formulation is neither correct nor incorrect, neither adequate nor inadequate, because it does not make empirical claims about existing seed plants. Empirical claims are only made by theoretical hypotheses that pronounce particular plants or groups of plants as instances of this model or theory.

The advantage of the semantic view of theories is obvious with regard to the classical theory. Seed plants that do not fit the classical theory do not pose a problem to the theory; they only limit its domain of applicability. Since a great deal of structural diversity in seed plants fits the classical theory, the latter performs a useful function.

The continuum model can also be understood in a semantic way which again means that it does not refer to actual plants. It has an advantage over the classical model because of its increased domain of applicability. In other words: it is more comprehensive; hence plants that do not fit the classical model are instances of the continuum model. In spite of this advantage, the classical model may still be used for all those many plants and plant structures that are instances of it. One reason for its continued use may be its relative simplicity (at least in the minds of botanists who prefer thinking in terms of mutually exclusive categories and hierarchies).

As pointed out above in terms of the traditional view of theories, the classical and the continuum models (or theories) might be seen as complementary to each other. The semantic view of theories makes it even easier to accept the coexistence of these two models. In this sense it has a liberating effect because it detracts attention from the notion of truth and adequacy of the theories, and focuses instead on the question of their applicability.

## To Chapter 2: Laws, Explanation, Prediction, and Understanding

*Laws.* One can consider laws as well-conformed general propositions within the framework of a theory. Thus, if the classical model is treated as a theory, the generalization that branching is axillary might be held to be a law. There is, however, the problem of invariance and universality. Laws in the strict sense (i.e., deterministic laws) are universal and therefore do not tolerate exceptions. In this regard, axillary branching is not a law. We know of genera in which branches are formed either on leaves or stems remote from the axil (see, e.g., Dickinson 1978; Guédès 1979).

It seems that there is a continuum from universal laws (if they exist) via weak laws to rules and finally irregular occurrences. If the generalization on axillary branching is still considered to be at the fringe of a weak law bordering at a rule, then it would be categorized as a coexistence law (or rule). This is evident if it is formulated as follows: in seed plants branching tends to occur in the axil of leaves. In other words: the properties of leaves along a stem and axillary branching coexist. An example of a process law is the statement that "in higher plants the operation of the physiologically active form of phytochrome ($P_{fr}$) leads to a stimulation of leaf growth while growth of the internodes is inhibited" (Mohr 1982, p. 101).

The weak law or rule of axillary branching is an empirical law or rule. Such laws or rules can be generated inductively. The observation of axillary branching in one plant or a number of plants can be genralized inductively. Since the law is not universal, one cannot, however, claim that all seed plants have axillary branching. One can at best conclude that this applies to most of them.

An example of a more general law (principle) that cannot be derived inductively is the principle of varying proportions. This principle allows the explanation of much of the diversity of plant form as the result of varying proportions. In classical morphology it may refer to varying proportions of organ systems, organs and subunits of organs such as the parts of leaves (Troll 1949). Thus it is based on the notion of the classical kinds of organs and therefore is subject to the criticism of the classical model. However, this principle can also be applied in a wider context (see, e.g., Thompson 1917; Bookstein 1978).

*Law and order.* I can now use the postulate of axillary branching as an example for the discussion of the difficult question of whether there are laws and order in nature. As pointed out already, we have found exceptions to the common case in which branching occurs in the leaf axil. Hence, the postulate of axillary branching cannot be considered a deterministic (= universal) law. As far as I know, the same conclusion can be drawn with regard to all other generalizations of comparative morphology, if not biology as a whole (see, e.g., McIntosh 1980/1982, pp. 39–40). With regard to the postulate of axillary branching, the question, then, is whether it is a probabilistic law. The answer to this question may be yes or no, depending on the point of view taken. Thus, with reference to the whole flora or vegetation on the earth, the probability that we shall encounter axillary branching in the vegetative plant region is very high. However, with regard to some genera and species such as *Phyllonoma integerrima*, the probability of finding exclusively axillary branching is very low since epiphyllous inflorescences occur regularly. In spite of these two differing viewpoints, one cannot deny that in the vegetative region of seed plants branching tends to be axillary. If this tendency is interpreted as order, then we have here a case of order in nature. This order is, however, not a rigid pattern. It allows for exceptions, which reflect two basic features of life, namely variability and individuality (see also Elsasser 1975, 1981; and Chap. 9). Exceptions, if taken seriously, point to the capricious or chaotic aspect of nature. Nature is, however, neither totally chaotic, nor rigidly ordered; she is the Unnamable.

Returning now to the example of axillary branching, the matter becomes more complicated in the reproductive plant region. Since in many taxa stamens cannot be clearly assigned to the categories of caulome or phyllome, the question of whether they are axillary branches cannot be well posed. Many female structures of flowers, i.e., the classical carples, present a similar problem. If the placentae that bear the ovules are interpreted as organ homologs, then all placentae inserted on the carples instead of in their axil are exceptions to the postulate of axillary branching. There are still other problems most of which are related to difficulties in the classical interpretation of flowers (see, e.g., Sattler 1973, 1974c). In terms of the semantic view of theories, these difficulties are to decide whether so-called difficult floral structures are instances of the classical model (or theory). One advantage of the continuum model is that this dilemma may nor arise.

*Explanation.* After this discussion of law and order in morphology, I can now turn to the problem of explanation. According to the covering-law model of explanation, either deterministic or probabilistic laws are required for explanation. To the extent that such laws are questionable, our explanations are questionable, although they may seem to fit the covering-law model. Figure 5.5 illustrates one of many such situations (see also Fig. 5.6). If the axillary branching postulate is considered a probabilistic law, we may apply the covering-law model: the covering law is then a probabilistic law and the resulting explanation a probabilistic explanation. If, on the other hand, the axillary branching postulate is interpreted as a rule, the covering-law model does not apply in the strict sense and therefore an explanation in terms of this model cannot be given. However, explanation as analysis and relating (Woodger 1967) can be achieved: structures that are puzzling are analyzed and related to others within the continuum model, and are explained in this fashion. This example shows that 'explanation' and 'description' are only relative terms: what according to the authors of the covering-law model may be only description, constitutes explanation according to Woodger.

| Probabilistic law (?): | Side-branches (incl. flowers or inflorescences) are initiated in the axil of leaves s. lat. | } Explanans |
|---|---|---|
| Pertinent particular circumstance: | In *Helwingia japonica* the developing inflorescence is displaced from the leaf axil onto the leaf (see Fig. 5.6) | |

| | The inflorescence of *Helwingia japonica* is inserted on the leaf (Fig. 5.6) | } Explanandum |
|---|---|---|

**Fig. 5.5.** Example of a debatable probabilistic explanation following the covering-law model (see also Fig. 5.6)

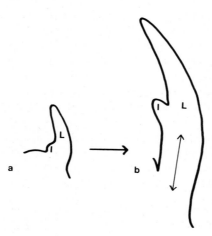

**Fig. 5.6a,b.** Diagrammatic longitudinal sections of developing leaves of *Helwingia japonica* (Thunb.) Dietr. in which the inflorescence (*I*) is initiated at the base of the leaf (*L*) in its axil (**a**). Due to an intercalary meristem, i.e., a growing region below the insertion of the inflorescence primordium (*double arrow*), the latter becomes displaced onto the leaf (**b**). (For microphotographs see Dickinson and Sattler 1975). Note that this species differs greatly from *Phyllonoma* (Figs. 5.2 and 5.3) whose epiphyllous inflorescence is not initiated in the leaf axil

*Prediction.* If the postulate of axillary branching would constitute a deterministic law, then we could predict for all plants that their side-branches are initiated in the leaf axil. However, since exceptions such as *Phyllonoma* occur, such prediction works only to a limited extent: a situation that seems to be rather characteristic of biology at the present time. In some cases exceptions may be so rare that for all practical purposes one may refer to a law and make successful predictions. Biologists who tend toward a lawful conception of nature emphasize those cases and may be eager to invent ad hoc hypotheses for the few exceptions. In other cases exceptions may be so

**Fig. 5.7.** Celebration: The rose (by Martin Carey). Ink on paper. 18 × 30″, 1967. This psychedelic drawing illustrates beautifully the oneness of rose and environment. [Reproduced from Masters and Houston (1968, p. 65)]

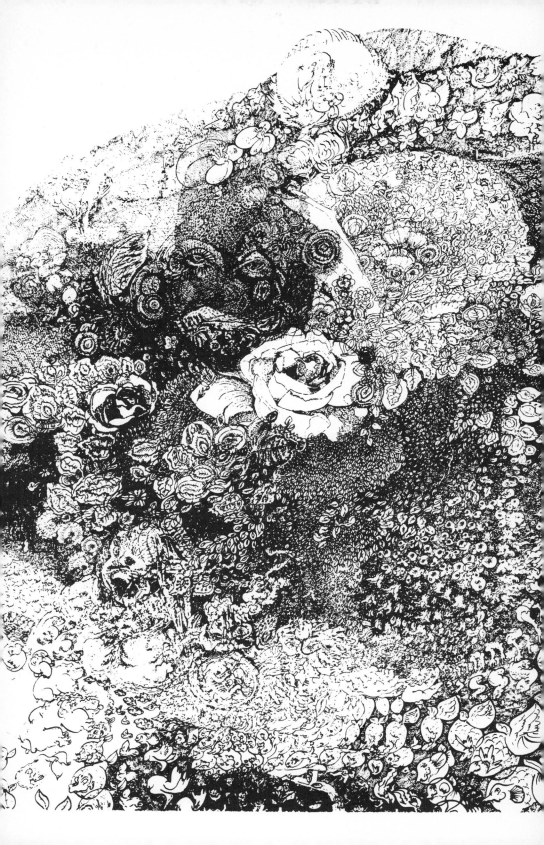

116

numerous that it is difficult to refer to laws or even rules. Biologists who tend to believe that nature is capricious stress those instances. I think that one has to acknowledge both aspects.

Since weak laws, rules, and even debatable rules, may allow us to make many correct predictions, predictability is no guarantee that there are laws in nature. Thus, the enormous number of correct predictions that may be made on the basis of the proposition that branching is axillary do not indicate that this proposition is (universally) correct. Any proposition that corresponds with reality to some extent is likely to produce many correct predictions. However, it will also lead to wrong predictions in cases that do not sufficiently agree with the proposition.

*Understanding.* As pointed out in Chap. 2, 'understanding' may have many meanings. With regard to plant morphology, this has been discussed by Arber (1950, 1954). Inasmuch as understanding transcends science and discursive thought, examples of such understanding cannot be illustrated in an objective, conceptual, and rational manner (Arber 1957). Fine arts (Fig. 5.7) and poetry (see Russell 1975) are better suited in this respect:

"Our lives are like the plants
floating along the water's edge
Illumined by the moon".
(Ryokan)
(translated by Stevens 1977, p. 77)

**To Chapter 3: Facts**

The statement 'this is a Magnolia flower' refers to a particularity and therefore constitutes a singular proposition or a fact. According to the first definition of fact, this statement describes something that exists independently of our observation, perception, and description of it. Specifically, this means that this Magnolia flower exists in nature as a separate entity that is composed of other entities, namely the floral axis (caulome) and the appendages (phyllomes). According to the second definition of fact, the above statement is a proposition of an objective datum of perception. It is objective because a consensus exists among present day botanists that 'this is a Magnolia flower.' One advantage of this definition is that its validity can be checked through an account of the perception of botanists. Another advantage is that it makes clear the requirements for a "fact": (1) we need a conceptual framework and a language to describe it, and (2) we need a consensus that a particular language and the concepts of "flower" and "Magnolia" should be used for the above description. We may ask illuminating questions such as the following: do all botanists agree on the use of the concepts "Magnolia" and "flower"? The answer is: no. At least some favor another description. For example, according to Melville (1962/1963, 1983), the Magnolia structure is a gonophyll of gonophylls which is quite different from a classical flower because it consists not only of one axis but of several and has a different architecture. Melville's description is not generally accepted. The other description in terms of the classical flower concept has the support of the majority of botanists. Although there is no complete consensus, i.e., no complete objectivity, we have a close approximation of a consensus. If we assume now that Melville's or other unconventional descriptions such as those by Croizat (1960, 1962; see also Heads 1984) or Meeuse (1981) were to gain ground and would gradually replace the classical one, we would pass through a phase in which neither one would be objective until the new one would become objective as a new consensus would be approached.

Facts and objectivity are relative because they are dependent on a conceptual framework which may be part of a theory and in that sense they are theory-laden. In the above example of the Magnolia flower the implied theories are either the classical theory or Melville's gonophyll theory. As the theory changes, the fact changes. Hence, a total independence of facts and theory seems illusory. Some facts may be more resistant to theory change than others, but this difference seems to be only one of degree.

Inasmuch as the concepts used for the description of facts depend on the cultural tradition, facts are also culture-dependent. Thus the term 'flower' can be used in the sense of blossom. In this sense it has been used in different cultures long before the classical theory had been invented. This shows that the concept "blossom" is rather robust toward cultural differences. It is possible, however, that even for the concept of "blossom" nuances of different meanings exist in different cultures. The importance of transcultural perspectives and investigations is evident at this point.

Cultural tradition may not only influence the conceptual framework that is used for the description of observations; it also relates to the state of consciousness which in turn may affect facts. For example, in our ordinary state of consciousness it is a fact that plants do not have auras. However, in altered states auras may be observed (see, e.g., Davis and Lane 1978), i.e., auras become facts for a community of people living in an altered state. If our whole society would develop an altered state of consciousness, auras would become facts (i.e., objective) according to our second definition.

Consensus does not necessarily constitute truth. The masses can be wrong just as individuals may be wrong. Conversely, the fact that only few people see auras does not necessarily mean that therefore auras do not exist. The advantage of the second definition of fact, which is based on the second definition of objectivity, is a distinction between consensus and truth. Both subjective and objective experiences may be true or false. Hence, facts are not necessarily real just because they are objective in terms of inter-subjectivity (i.e., in terms of the second definition).

Do particularities exist as such? A flower is continuous with the remainder of the plant. A whole plant is continuous with the environment. Thus a flower exists not by itself but in the context of the whole. In that sense a flower and anything that is said about it do not constitute an ultimate reality because reality does not just occur in fragments. Fragments, i.e., particularities, are our creations. Yet, although they are our creations, I cannot imagine that they are totally arbitrary creations. It seems that facts are constrained by reality, although they may be arbitrary to a certain extent.

With regard to the flower example the primary realm is a patch of colors, shapes, etc. in its context. Instead of trying to describe it by words, the primary realm might be better conveyed through a picture of a flower in its context because the words 'color' and 'shape' introduce already concepts that are not present in the primary realm. A picture such as a drawing or painting is, however, dependent on the state of consciousness of the artist. Thus, a psychedelic artist draws a rose differently from an artist who is not under the influence of drugs (Fig. 5.7). Furthermore, there are other factors, such as the technique utilized, that influence the way a flower is pictured.

The primary realm can neither be pictured nor described conceptually, it must be experienced in a state of mind that transcends discursive thought. Words, which usually refer to concepts, are a hindrance in the communication of direct experience. Hence, the sage who can see reality in its suchness is often silent. Zen poetry is an attempt to say through words what cannot be said by words. Thus, Ryokan who led a reclusive life in celebration with nature wrote (see Stevens 1977, p. 59):

> Spring flows gently –
> the plum trees have bloomed.
> Now the petals fall, mingling with the song
> of an uguisu.

## To Chapter 4: Concepts and Classification

*Concepts.* "Leaf" is a concept that may be formed through the abstraction of the following features: (1) dorsiventral symmetry, (2) determinate growth, (3) lateral position on the stem. If other features are added, such as, for example, the presence of an axillary bud, we arrive at a different concept of "leaf" because the definition is different. Theoretically, it could happen that both concepts have the same extension. Yet, in this instance this is not the case because there are leaves with and without axillary buds.

Many students fail to see that "leaf" is a concept. They assume that "leaf" is something concrete because one can point at it. Yet one can point at a leaf only after one has drawn an imaginary line between the "leaf" and the "stem." In other words: one can speak of a leaf only after one has separated it from the remainder of the plant through the process of abstraction. Actually, there is a continuity between the "stem" and the "leaf" (see, e.g., Howard 1974; Rutishauser and Sattler 1985). Hence, a "leaf" as an entity does not exist in nature, but is a thought-object. The question is whether this thought-object possesses a high or a low degree of adequacy. In terms of Woodger's map analogy one can say that the concept of a so-called natural entity portrays nature relatively well, whereas the concept of an artificial entity shows a rather low degree of correspondence between the concept and nature. Conceptual realists assume, however, that natural and artificial concepts are mutually exclusive: the former are real, i.e., exist in nature and the latter represent only a mental construct with no reality in nature. Yet even the most natural concept results from abstraction and therefore does not completely correspond with nature. It portrays at best one aspect of nature. On the other hand, an artificial concept also entails a correspondence with nature, although to a very limited extent. Hence, the difference between a natural and an artificial concept is relative. It might be better to drop those terms in order to avoid misunderstandings.

"Leaf" ("phyllome") and "stem" ("caulome") have been considered real (Troll 1937). However, their continuity and the occurrence of intermediates does not support their reality as mutually exclusive essences (Sattler 1974; Sattler, Rutishauser and Luckert, in press). Nonetheless, the concepts "leaf" and "stem" are not artificial because typical leaves and stems are relatively articulated and are quite frequent in contrast to the relatively rare intermediates that dissolve the discreteness of the classes "leaf" and "stem".

*Domain of applicability.* Traditionally the concept of "carpel" is defined as a phyllome that bears and encloses ovule(s). This concept has sometimes been applied beyond its limits of applicability, i.e., to a domain that does not satisfy the definition of the concept. For example, when the ovules are born on the floral axis (caulome), no ovule-bearing phyllome exists. Hence, the above definition does not apply. To deal with such cases, one could redefine the term 'carpel' as follows: 'carpel' is a phyllome that encloses ovule(s) (Sattler and Perlin 1982). According to this definition it is not required that the ovules are inserted on the phyllome. Therefore, the case in which the ovules are born on the axis satisfies this different 'carpel' definition. Examples of carpels in the latter sense occur in many groups of flowering plants (see, e.g., Sattler 1973; Sattler and Perlin 1982).

*Term and concept.* The two definitions of 'carpel' are an example in which the same term, namely 'carpel', designates two different concepts. On the other hand, 'folded megasporophyll' and 'carpel' (in the sense of the first definition) are two different terms for the same concept.

*Definition.* Definition, although important, never can be complete. It is always based on terms that must remain undefined and thus can only be grasped intuitively. For example, 'carpel' is defined in terms of 'ovule' and 'phyllome.' 'Phyllome' is defined in terms of 'laterality,' 'dorsiventrality,' and 'determinate growth.' The latter terms are again defined through other terms, and so on, until we arrive at primitive terms that remain undefined.

*Operationism.* The original notion of operationism in the sense of Bridgman is not easily applied to plant morphology. However, diluted versions of operationism such as the one referred to by Sneath and Sokal (1973, p. 17) can be used in plant morphology. Thus, terms such as 'leaf' may be defined operationally in terms of the three or four properties listed above. Any generalization that contains the operationally defined term 'leaf' may then be subjected to empirical tests.

*Polytypic versus monotypic concepts.* The concept of "leaf" mentioned above is a monotypic concept defined by a set of properties. According to this definition, a structure is a leaf only if it possesses all the properties required by the monotypic concept. When we encounter a structure that intuitively appears to be a leaf but lacks one of the defining properties, this structure must be excluded from the concept of "leaf." Intuitively we may consider such exclusion unnatural because we feel that the excluded structure belongs to the leaf category. The notion of polytypic classes may provide a logical underpinning for this feeling. According to a polytypic concept of "leaf," a structure must only have a large number of the properties of the set, and each of the properties must be possessed by a large number of the structure belonging to the polytypic class. If we consider three out of four large, a structure that lacks axillary buds may still satisfy our polytypic leaf definition in terms of four properties if it exhibits the remaining three properties of laterality, dorsiventrality, and determinate growth. Thus a polytypic leaf concept is more adequate than a monotypic one because it avoids the artificial exclusion of structures that lack one or the other property.

*Fuzzy sets.* Another problem arises when we encounter structures that are intermediate between leaves and stems (e.g., the phylloclades referred to above). Such structures fit neither a monotypic nor a polytypic concept of "leaf" or "stem." In this situation the fuzzy set notion is more adequate.

According to this notion, structures may have 0% to 100% membership in the fuzzy set of "leaf." Thus, a structure intermediate between a stem and a leaf may have a 50% membership in the fuzzy leaf set. At the same time this structure may have a 50% membership in the fuzzy set of "stem." Fuzzy-set theory allows membership in different sets and thus shows various relations of one structure. According to monotypic and polytypic definitions a structure must belong to *either* one *or* another class. In contrast fuzzy-set theory may show the network character of relations.

*Injunctions.* Since it is difficult to quantify class membership, it may be best at the present time to treat class concepts such as stem and leaf as injunctions. This means that they are not defined monotypically or polytypically. The fact that most morphologists have actually avoided defining leaf, stem, and other concepts might indicate their intuitive feeling that a definition is impossible. Nonetheless, it seems that they are not treating these concepts as injunctions because that would imply that certain structures could be partial members of the leaf and stem concepts. Usually this consequence is not accepted.

*Types.* Types can be considered as monotypic or polytypic classes, or even as fuzzy sets and injunctions. The latter definition appears most appropriate, especially in cases of great plasticity and variability as, for example, in stems and leaves.

*Conclusions.* Although polytypic concepts, fuzzy sets and injunctions allow a more adequate description in many instances, they still involve abstraction. Abstraction removes us to some extent from reality, creates fragmentation and obscures the unity. Thus the leaves, stems, and roots of one plant are continuous with each other and form a unity. The plant itself forms a unity with the environment. A direct experience of this unity requires the transcendence of the conceptual approach.

Things, forms, and classes do not exist as such in nature. Yet they are not totally fictitious. Inasmuch as they are adequate they may be a useful guide for living. Thus a knowledge of the class of poisonous plants may be crucial for survival.

*Classification.* The distinction of classes is already classification in the wide sense. Thus, for example, the distinction of the classes phyllome and caulome constitutes a classification. However, classification in the strict sense refers to the construction of a system of classification. As an example of such a system I shall not use the customary taxonomic system, but the following morphological classes in their traditional hierarchical relation:

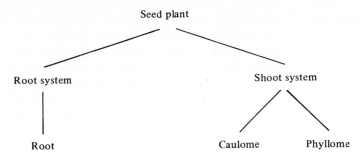

The classes root, caulome, and phyllome that comprise the lowest level of the above (system of) classification are not given, but have to be established in terms of other concepts such as symmetry, growth pattern, etc. It is debatable to what extent they are adequate classes.

In the relatively simple example of the above morphological (system of) classification not many possibilities of grouping exist. One possibility is to combine caulome and phyllome to form the higher-level class of the shoot system, whereas the class of the root system comprises only one subclass, namely the root class. An alternative would be to unite the classes root and caulome into the class of the axial system in contrast to the class of appendicular organs (i.e., phyllomes) which form the appendicular system:

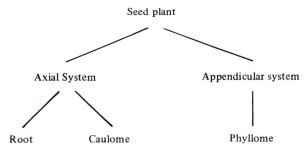

Which of the two morphological systems is chosen may depend on the weighting of characters. If, for example, the symmetry character is given special weight, the classification into an axial and appendicular system may result. Still other classifications result, if plants are subdivided into different units due to different abstraction. Thus, for example, the shoot may be thought to consist of one type of growth units (phytons) only (see Rutishauser and Sattler 1985).

Morphological systems like taxonomic systems may be interpreted in terms of similarity or evolution. Furthermore, it can be debated whether hierarchical or net-systems are more adequate. In my opinion, the affinities between organs are more adequately described by a net-system. The continuum model of the shoot that I proposed is a simple net-system (see Fig. 5.4a). The space between the four extreme concepts of shoot, caulome, phyllome, and trichome is filled with intermediates (for examples see Sattler 1974). Each of the four concepts functions as an injunction or fuzzy set so that any individual structure that occupies the space between the four points can be a partial member of two, three, or four fuzzy sets. For example, structures occurring on the leaves of *Begonia hispida* var. *cucullifera* may belong 100% to 0% to the leaf class and 0% to 100% to the trichome (or emergence) class. In other words: some are leaves, others are trichomes, and still others are intermediate between leaves and trichomes (Sattler and Maier 1977).

## Conclusions

Many modern biologists tend to find the whole field of comparative morphology either uninteresting or outdated, which means that it is no longer at the forefront of basic and challenging biological research. I hope that the preceding discussion has shown that this is not the case. Like any field, comparative morphology comprises classical and modern approaches. The classical approach remains useful for much "puzzle solving" in terms of normal science of the classical paradigm. Alternative approaches such as the continuum approach change the core of fundamental classical postulates, such as the discrete (atomistic) and hierarchical organization of plant construction. Possibilities for further innovations abound. Thus, comparative morphology, like other disciplines, is open to major (r)evolutions. In fact, it is no exaggeration to say that it awaits the great genius who will be able to propose an attractive and elegant theory that is highly adequate, or, in terms of the semantic view of theories, extremely comprehensive.

Comparative morphology deals with the comparison of structures and forms. This means it is concerned with two fundamental features of living systems: diversity and structure. In contrast, much modern research in biology is focused on the detailed analysis of a few selected organisms that are suitable and convenient for this type of research. Such specialization need not be negative, provided one keeps in mind that life is diverse. Comparative studies are, however, important to provide a balanced overall perspective.

With regard to structure and form, it is now more evident than ever before in the history of biology that these features are characteristic of all levels of organization. As White (1968, p. XI) noted, "this is a world of form and structure and can only be properly understood as such." It is, however, important to realize that structures and forms are constantly changing during ontogeny and phylogeny (see Chap. 8 on Evolution and Change). Therefore, a static approach, which to some extent is characteristic of much classical morphology, has limited adequacy and usefulness. A dynamic approach that focuses on structural transformation will give a more adequate representation of nature (see, e.g., Whitehead 1929; Sattler 1984). A corollary of this approach is the dissolution of the form-function antithesis in at least one sense (see, e.g., Woodger 1967; and Chap. 7 on Teleology).

Some biologists continue to argue that the most important levels of organization are the cellular and molecular. The inadequacy of this view will be discussed in Chap. 9. Due to the phenomenon of emergence (see Chap. 9), organization at the levels of organs and the whole organism cannot be reduced to the cellular and molecular levels. As de Bary noted long ago, "the plant forms cells, not cells the plant" (see Barlow 1982). Additional criticism of cell theory (see Hagemann 1982, Chap. 2) further weakens and limits cell-biological approaches.

Finally, it has been argued that the top challenge of modern biology is the causal analysis of living systems. As pointed out by Mohr (1977) (see Chap. 6), instead of causal analysis we can only accomplish factor analysis. Inasmuch as factor analysis is a search for correlations, it approaches the methodology of comparative morphology which is also concerned with correlations. Thus, the difference between the factor analysis of the physiologist and the methodology of the comparative morphologists need not be as fundamental as often assumed. However, even if there is a basic difference in methodology and research goals of physiologists and comparative morphologists, one cannot assign priority to the physiological approach, one reason being the dependence of physiology on comparative morphology. In physiology morphological terms such as, for example, leaf, stem, and flower, are indispensable. Hence, to the extent that physiological statements utilize morphological terms, they reflect morphological thinking. Inadequacies in morphology may thus be introduced into physiological factor analyses and may even influence the formulation of research questions and projects, including the experimental design. Progress of comparative morphology, whatever "progress" may mean, is at least to some extent basic to progress in the factor analyses of morphogenesis, physiology, genetics, ecology, and other fields (see, e.g., Wardlaw 1965; Steeves and Sussex 1972; Stebbins 1974; Hallé et al. 1978; White 1979, 1984; Tomlinson 1983; Gottlieb 1984; Barlow and Carr 1984; Dickinson and White 1984; Malacinski and Bryant 1984; Cusset 1985, 1986).

# 6 Causality, Determinism, and Free Will

> "The principle of causality is never strictly true because it neglects two ever-present features: spontaneity (or self-determination) and chance" (Bunge 1979a, p. XXII)
> "The reason why physics has ceased to look for causes is that, in fact, there are no such things. The law of causality, I believe, ... is a relic of a bygone age."
> (Bertrand Russell 1913, p. 1)

## 6.1 Causality

### 6.1.1 Introduction

According to my experience, the two most common kinds of questions biology students ask when encountering a new structure or process are the following: What is the cause? What is the purpose (or function)? An answer to the two questions is often considered tantamount to an understanding of the phenomena. In this chapter, I shall examine the notion of "cause" and "causality," whereas purpose and function shall be discussed in the next chapter.

Causality is a controversial subject in the physical as well as the biological sciences (see, e.g., Brand 1979; Salmon 1980). Many biologists take it for granted that everything must be explained or must be explainable by a cause (or causes). Critical biologists realize, however, that "whether the principle of causality is true is a difficult question to decide" (Mohr 1977, p. 76). In spite of this difficulty and the resulting ambiguity concerning the general validity of causality, many biologists insist that "the principle of causality is an indispensable prerequisite of science, irrespective of whether we classify it as a universal law or as a metaphysical postulate" (Mohr 1977, p. 77). Alternatively, there are biologists who think that the notion of causality is unnecessary or even undesirable for the conduct of research and the understanding of nature (Peters 1983). Intermediate standpoints exist, of course, according to which weaknesses of the notion of causality are admitted without demanding a total abolition of causal thinking. For example, Woodger (1967, p. 189) agrees with Mach that "there is no cause and effect in nature," yet he tends to acknowledge a certain usefulness of causal analysis of biological phenomena. He warns, however, that the notion of causality is beset with problems that need clarification (see Bunge 1979).

Like biologists, philosophers are divided in their views of causality. Many of them consider the notion of causality as indispensable or go as far as to assume that causality is real and universal. Plato (in his *Timaeus*) said: "Everything that is born, is necessarily born through the action of a cause" (see Caratini 1972, I-21). Aristotle (in his *Metaphysics*), also accepting causality, distinguished four kinds of causes: the efficient cause (causa efficiens), the material cause (causa materialis), the formal cause (causa formalis), and the final cause (causa finalis). The difference between these four causes is usually illustrated by the example of the construction of a house. The efficient cause propels the construction in terms of energy, money, or construction workers. The material cause constitutes the necessary building materials. The formal cause is the plan of the architect which determines the form of house. The final cause is the goal of the person(s) without whose intention the house never would have been built. In modern science, including biology, "cause" tends to be used in the sense of "efficient cause." "Final cause," if considered at all, is treated in the context of teleology (see the next chapter on Teleology).

In the following discussion of causality, I shall follow the general practice of scientists and philosophers of science to use the term causality only in the sense of "efficient causality." How has this term been traditionally defined? Needless to say, different definitions exist (see, e.g., Bunge 1979a). According to the *Encyclopédie thématique universelle* (Caratini 1972, I-21), causality is the relation of cause and effect, whereby A is the cause of B and B the effect of A if and only if the following three conditions apply:

(1) B follows A in time, i.e. A and B form a succession
(2) A produces B
(3) The relation between A and B is constant and necessary.

In terms of this definition of causality, we may consider the standpoints of other philosophers who have made major contributions to the evaluation of causality. Hume rejected this idea of causality because he did not accept the second condition and the postulate of necessity in the third condition. He acknowledged that we observe regular successions of events, such as, for example, the succession of day and night or the succession of the rising of the sun and the rising of the temperature. However, Hume insisted that this is all we can experience. We do not observe that A actually *produces* B, by necessity. Thus, according to Hume the idea of necessary production transcends experience; its acceptance would amount to a belief.

Kant agreed with Hume that causality is not just a matter of experience. It is a prerequisite, a category of experience without which experience is impossible. Hence, according to Kant, we have no choice but to accept it even if reality itself may not be causal at all.

Auguste Comte and following him the positivist and neopositivist thinkers denied causality altogether. Bertrand Russell (1913, p. 1) wrote that "causality is the relic of a bygone age surviving ... only because it is erroneously supposed to do no harm." Wittgenstein (1922: 5.1316) claimed that the "belief in the causal nexus is superstition." Collingwood (1940) stressed the relativity of causes, i.e., causes are relative to the point of view adopted. "For a mere spectator there are no causes" (Collingwood 1940, quoted by Dray 1964, p. 46). Kuhn (1977) discussed the general decline in causal thinking in physics: "though the narrow concept of cause was a vital part of the physics of the seventeenth and eighteenth centuries, its importance declined in the nineteenth and has almost vanished in the twentieth" (Kuhn 1977, p. 28).

It would be one-sided to quote only those modern philosophers who are either very critical of causality or who deny its existence and usefulness altogether. There are also modern philosophers who claim that the notion of causality is either totally or partially appropriate to deal with nature (e.g., Mackie 1975). Thus the controversy over causality continues. Whereas some think that it is "a relic of a bygone age" (Russell), others see a comeback of causal thinking, especially in the biological, psychological, and social sciences (Bunge 1979a, p. XVII).

After this brief sketch of some of the many points of view of philosophers, I shall now turn to more concrete biological considerations, which may be more meaningful for students of biology. I shall attempt to show that developments in biology have led to a clearer view of causality and thus contributed to our knowledge of biophilosophy. This indicates that science may also help to solve general philosophical problems, although it cannot free itself completely of its philosophical foundation (see Introduction). Science and philosophy are interacting with each other and in this way are mutually dependent on each other (see Wuketits 1978).

## 6.1.2 Linear Causality

In biology we often enquire about the cause of a phenomenon. When we think that we have located the cause, we want to find out the cause of the cause, and so forth. In this way, we hope to establish cause after cause until, so it is said at times, we shall arrive at the first cause, the ultimate cause that would explain everything. The notion of the first or ultimate cause is, however, a very spurious one, which I do not want to pursue any further for reasons given below. What matters at this point is that this common practice of biologists to look for causes of the causes may lead to what is called a causal chain:

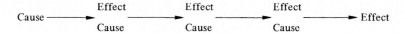

In this chain each cause has an effect that in turn is the cause of another effect, and so on. Linear thinking in terms of such chains is still employed in biology as well as in everyday life. For example, the fact that a driver was daydreaming may be considered the cause of an accident, which in turn may be the cause of his injury, which finally may be the cause of his death. Chains of this sort are linear connections of events or things; hence the term 'linear causality.'

Often, an effect is seen as the result of more than one cause. We speak then of multiple causation by a plurality of causes (e.g., Wolvekamp 1982; Hilborn and Stearns 1982). In such cases the cause-effect relations have a branched or hierarchical form as follows:

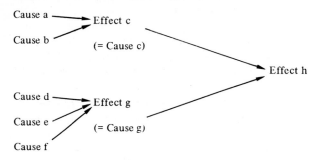

In contrast to the above scheme, which illustrates the plurality of causes, one single cause may be said to have a diversity of effects. This again may lead to a hierarchical scheme. Although this way of hierarchical interpretation may allow us to deal with connections that cannot be satisfactorily analyzed by simple (i.e., unbranched) causal chains, it is not yet sufficiently adequate to deal with the complexities of living systems. Chains and hierarchies form loops and networks. This leads to the notions of circular and network causality, which are no longer causality in the traditional sense as defined by the three conditions above (Sect. 6.1.1).

## 6.1.3 Circular Causality

Many examples of feedback have become known in biology. In these cases the effect of the cause has an effect on the cause by which it is produced:

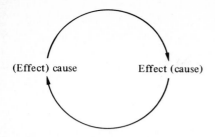

(Effect) cause          Effect (cause)

Strictly speaking, "in the feedback loop there is no meaning in establishing cause and effect, because cause and effect are mixed together" (Varela 1976). Thus, the notions of cause and effect lose their original meaning in this situation.

Traditionally, cause and effect are defined in a temporal sequence (see the definition above, Sect. 6.1.1), i.e., the cause precedes the effect in time. In the case of circular causality this is no longer true because the circle does not necessarily close itself in time, but it moves through time as a circle. This means that the notion of circular causality is quite different from the traditional notion of causality as it was presented in the above definition (Sect. 6.1.1) and in the scheme on linear causality. To use the general term 'causality' for the notion of "circular causality" may suggest a similarity that does not exist. Circular causality is not just a subcategory of causality, but a concept that supersedes the traditional notions of cause and effect. Hence, these traditional notions no longer apply.

### 6.1.4  Network Causality

Modern biology has shown that the notion of circular interactions is also a simplification, although a much more useful one than linear causality. The circles of circular causality also interact with each other and thus a more adequate representation of interactions in living systems is by means of a network (for an example see Fig. 6.1; for other examples see Bindra 1976, p. 42; McMurray 1977, p. 278; Lewis 1977; Vester 1978; MacDonald 1983). From this point of view causality is conceived as network causality. As in the case of circular causality, network causality does not imply the traditional notion of cause and effect. Thus, network causality is also a concept quite different from the traditional concept of causality; hence, for the sake of clarity the former might be given a name that does not include the term 'causality.' Network causality is much more akin to circular causality than either one of them is to linear causality, because both network and circular causality are in opposition to the basic assumptions of traditional causality, namely the relation of cause and effect as postulated in the definition above

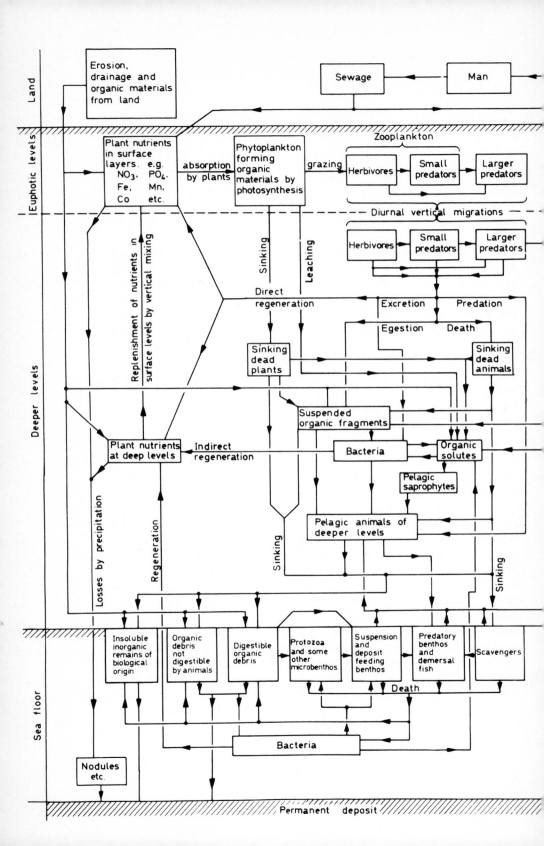

(Sect. 6.1.1). Because of this fundamental difference, certain authors no longer consider network causality as causality. Hence, they conclude that causality has outlived its usefulness. Other authors, however, consider network causality simply as a different kind of causality. Consequently, these authors conclude that causal thinking is still fundamental to modern biology, although the notion of causality has changed fundamentally. Obviously, the two conclusions, which seem to be contradictory, are only two semantic versions of the same postulate, namely that our traditional causal thinking has undergone an enormous change. Whether the new way of network thinking is still termed causal or given another name is a matter of convention. I would prefer a new name in order to underline the difference of network thinking. Perhaps the terms 'network interactions' and 'network thinking' would be appropriate. I shall, however, use the term 'network causality' in certain cases simply as a matter of convenience.

Although it is probably generally recognized that network causality constitutes a more adequate and comprehensive representation of the interactions in living systems, it does not necessarily follow that therefore circular and linear causality have become totally useless concepts. Circular and even linear causal connections may present at least a certain segment of the net of interactions and in this way provide some limited information (see Bunge 1979a). Furthermore, if we assume that some of the interactions in the network are strong, whereas others are rather weak, then a chain, a hierarchy or a circle may actually represent a relatively autonomous segment of nature. In the diagram of Fig. 6.2 strong interactions are indicated by solid lines and

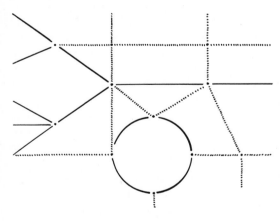

**Fig. 6.2.** Hypothetical network of strong (*solid lines*) and weak (*dotted lines*) interactions. The strong interactions provide a basis for linear, hierarchical, and circular causality, whereas the totality of strong and weak interactions represent network causality

**Fig. 6.1.** Some interconnections of the marine food web (After Tait 1981, p. 204; reproduced by permission of the author)

weak ones are dotted. This illustrates a hierarchical and linear segment as well as a circular one (see also Wuketits 1981, p. 86). Since the use of the notions of linear and circular causality has been successful to a certain extent in terms of causal explanation and prediction, it seems that the kind of diagram of Fig. 6.2 is not totally fictitious. In other words: there seem to be differences in nature between strong and weak interactions in such a manner that they provide an empirical basis for the construction of causal chains, hierarchies, and circles. This means that the notions of linear and especially circular causality may be adequate to a certain extent within a limited context. To what extent they will remain useful will have to be demonstrated empirically by an analysis of the strength and weakness of interactions. However, even if future research should provide a relatively firm foundation for the use of the notions of linear and circular causality, a more complete and comprehensive representation of living systems will have to resort to network causality. It has become increasingly evident in the biological as well as the psychological and social sciences that the systems under investigation exhibit a high degree of complex and netted interactions that cannot be understood in terms of traditional notions of causality (see, e.g., McMurray 1977, p. 278; Bunge 1980, p. 42; Wuketits 1981).

One fundamental implication of network causality that also shows how radically it differs from linear causality is the following: any particular effect (i.e., thing or event) in the network is not caused by one or several causes, which traditionally are thought to be things or events occurring temporally before the effect. Any particular effect, which I prefer to simply call a particular thing or event, is determined by the whole network of interactions. It may have been "triggered" by a specific change in a specific place. If "trigger" is equated with "cause," then we could, of course, speak of "causation." However, such "causation" is quite different from the traditional meaning as described in the definition given above (Sect. 6.1.1). One difference is that a trigger does not produce an effect. The effect is produced by the trigger in conjunction with the whole network (or at least segments of the network). Another difference is that the same trigger may have quite different effects depending on differences in the network. Hence neither condition 2 nor 3 of the traditional definition of causality (see Sect. 6.1.1) apply to the notion of "trigger" (see also Weiss 1973, p. 99). If one prefers to define cause as "trigger," then a very weak notion of cause results. Causes in this sense "are effective solely to the extent to which they trigger, enhance, or dampen inner processes" (Bunge 1979a, p. 195). Such "causes" only contribute to the determination of the effect, but do not produce them entirely. This shows again that the term 'cause' has different meanings, which may be such that notions intermediate between linear, circular, and network causality may result.

Finally, a limitation of network thinking may be pointed out in terms of Bohm's (1980) theory (or paradigm) of the implicate order. According to this theory, the implicate order underlies the explicate order to which we are accustomed. Whereas the latter is diversified and fragmented (i.e., consisting of things and events), the former is "holomovement" which is "undivided wholeness in flowing movement" (Bohm 1980, p. 11). "This implies that flow is, in some sense, prior to that of the 'things' that can be seen to form and dissolve in this flow" (Bohm 1980, p. 11). "All parts of the universe, including the observer and his instruments, merge and unite in one totality. In this totality, the atomistic form of insight is a simplification and an abstraction, valid only in some limited context" (Bohm 1980, p. 11). This kind of thinking obviously goes even beyond network thinking because the points and lines in the network of the explicate order are no longer singled out in the implicate order. The latter underlines unity, wholeness, and total integration.

### 6.1.5 Mohr's Model of Factor Analysis

According to Mohr (1977), the way in which I used the term 'cause' in the preceding discussion is not quite correct. I shall therefore present Mohr's model of analysis and I shall show that it leads to the same general conclusions as in the preceding discussion of network causality. According to Mohr's model, biological causal research "must always be regarded as 'factor analysis,' rather than as causal analysis" (Mohr 1977, p. 77). All we can do in so-called causal research is "to vary one, two, or (rarely) more factors in an experiment and to record the effect. Those of the x factors that we vary in the experiment are called variable factors or briefly "variables" (Mohr 1977, p. 77) (see Fig. 6.3).

Since the cause a is determined by x factors, a causal analysis would require a knowledge of all x factors and how they determine the cause a and the

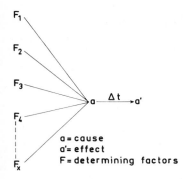

Fig. 6.3. Diagram showing how x factors ($F_1$ to $F_x$) determine the cause $a$, which over time ($t$) produces the effect $a'$. Only $F_1$, or $F_1$ and $F_2$, are the variable factors (After Mohr 1977, p. 77)

resulting effect a'. Mohr (1977, p. 78) points out that "even with the most simple living system such as bacterial cells, we are far from being able to perform a full-scale causal analysis. Therefore the physiologist will restrict himself to a factor analysis" (Mohr 1977, p. 78). In other words: only relatively few factors, namely the variables, are considered and in the factor analysis it is shown how these variables *influence* a particular feature of the system. It is not shown how they *cause* it, because the feature is not only caused by the variables, but also by all the other x factors.

It is a grave misunderstanding that many biologists who are used to working with variables tend to think that the variables cause the effect. This misunderstanding can have profound consequences and leads to distorted views of living systems such as, for example, the postulate of genetic determinism according to which the genes determine the traits of organisms [for an excellent criticism of genetic determinism see Weiss (1973), Lewontin et al. (1984)].

Mohr (1977) uses an example of a single factor analysis in genetics to illustrate the relation of factor analysis and causal analysis. This example concerns the factor that influences (we are tempted to say *causes*!) the formation of anthocyanin, a red or blue pigment, in many plant organs such as the petals of violets. At first Mohr equates the x factors (which according to his model cause the red pigmentation) with a set of x genes and points out that only one gene has been pinpointed as a variable, i.e., without this particular gene the pigmentation will not develop, although all other genes of the set x are present. On the basis of this experimental evidence it is often said that the red pigmentation is caused, i.e., produced, by the one gene. This kind of conclusion is fundamentally wrong because it mistakenly equates a single factor with the cause. In order to obtain the pigmentation the whole set of x genes is necessary. And furthermore, other factors of the organism and the environment are required. For example, the environmental factor light is necessary for the synthesis of the pigment. This means that the total set of x factors that causes the pigmentation does not only include the x genes, but must be extended to include the factors of the whole system of the organism and the environment of the organism. Thus, by this more rigorous form following Mohr's model of factor analysis as it relates to causation we arrive at the same conclusion as that reached in terms of network causality: any particular event such as the formation of a pigment is determined by the whole system, which includes the environment.

Figure 6.3 is a very simplified presentation of the total set of internal as well as external factors which determine a particular effect. The network between the factors is not indicated. Naturally, the cause should not be construed as having a distinct existence separately from the factors (see Woodger 1967, p. 189, footnote). Since the cause is constituted by the factors, we might just as well refer to the factors only. This would provide greater clarity of the situation. Another reason for eliminating the concept of cause has been mentioned in connection with the network model of interactions. If any particular thing or event is determined by everything else, i.e., the whole system, then the concept of cause is overextended. Evidently, the original meaning of "cause" no longer applies at the level of networks.

### 6.1.6 Consequences of Network Thinking

The consequences of the network view of multifactorial interactions are manifold and profound for science, society and our personal existence. In my opinion, this view may revolutionize science and society to an extent beyond our present imagination. Such a revolution can, however, only occur if we pursue the consequences of networks in all aspects of our lives. This is not at all easy because we have become so deeply conditioned by the linear way of thinking that we tend to slip back into its oversimplification and distortion without even noticing it. Some biologists disagree with this view. Thus, Riedl (1980) and other evolutionary epistemologists tell us that we think in a linear and hierarchical fashion because it is genetically determined and furthermore has a basis in nature. I think, however, that all we inherit is a certain potential (see also Weiss 1973; Lewontin et al. 1984) and that our actual linear and hierarchical thinking results from social conditioning. Having grown up as one of the victims of this social conditioning, but recognizing now the evidence for networks due to developments in modern biology and personal experiences, I shall attempt to present very briefly a few of the many consequences of networks and network thinking.

(1) As Wuketits (1978, 1981) and other systems-oriented thinkers amply demonstrated, networks imply structural and functional integration. This means that we have to look at organisms as wholes and furthermore to perceive them as integrated into the more inclusive ecosystem. The ecological movement is recognizing this, but far from drawing all the important consequences for ecology and society (see, e.g., Skolimowski 1981).

(2) The scientist as observer can no longer be excluded from the network(s) he studies because(s) he is part of it. Hence, it cannot be taken for granted that thoughts and emotions of the observer have no influence on the observed (see also Chap. 3).

(3) Since the whole is necessary for the understanding of the parts, investigations of the parts in isolation as is generally done, for example, in in vitro studies, can be of only limited usefulness for an understanding of the whole system and the role of the parts in that system. Mohr (1977, p. 81) is acutely aware of this problem when he writes with regard to plant

physiology: "any extrapolation from segment physiology to whole plant physiology must be considered with great caution. It is legitimate, of course, for a plant physiologist to analyze the growth regulators in the case of the isolated pea epicotyl or *Avena* coleoptile section; it is not legitimate, however, to assume as a matter of course that this undertaking will contribute to our knowledge about growth regulation in the intact system."

(4) Since network thinking concerns the very basis of biological research, it may profoundly affect the questions we will ask, the experiments we shall perform and design, and consequently, the direction in which future research will proceed.

(5) It is said at times that all biological explanations have to be causal explanations in order to be good explanations. According to the covering-law model of explanation (see Sect. 2.2), this statement implies that all biological laws have to be causal laws. If a causal law has the form: if cause a, then effect b, and if 'cause' is defined in the traditional sense, then causal laws are not tenable unless in a weak and limited form inasmuch as linear causality can be considered to be an acceptable approximation to nature (see above Sect. 6.1.4). In order to avoid the questionable component of causal laws, namely the notion of traditional causality, one has to convert causal laws into laws of correlation of the form: if a, then b (a and b being things or events). Such laws may provide better explanations in terms of the covering-law model. The events or things a and b may follow each other in time or may be simultaneous. In the latter case, the law would be a co-existence law (see Sect. 2.1).

(6) Evolutionary theory is often considered the most comprehensive theory of modern biology. Wuketits (1978, p. 146) claims that evolutionary theory in its generally accepted form, the Synthetic Theory of Evolution (STE), is based on the notion of linear causality. He points out how network causality coupled with systems theory will provide a more complete explanation of evolution by considering the interactions of the whole system in its external and internal aspects. This leads, according to Wuketits (1978), to a recognition of "internal selection" in addition to "external selection" and consequently to a clearer view of the interactions between the organism and its environment (see Sect. 8.5.2).

(7) With regard to society and our personal lives the consequences of networks may be even more far-reaching. Riedl (1980, p. 118) refers to a profound connection between the traditional notion of cause and that of guilt. He points out that in Greek the original meaning of cause is guilt. Thus, looking for the cause meant looking for the guilt. Since this way of thinking has influenced us, it seems possible that there is a connection between the prevalence of our traditional causal thinking and guilt feelings which are also reinforced by our culture. As we are used to looking for causes every-

where, so many of us search ubiquitously for guilt. Thus, either my wife or I must be guilty if our marriage broke down; or the thief must be guilty if (s)he stole our money. The realization that the search for guilt may be just as misguided as that for causes may profoundly affect society as well as the individual.

(8) The consequences of linear and hierarchical causal thinking may be even more tragic. Riedl (1980, p. 146), quoting Dörner (1975), points out that people with the best intentions of improving the world end up destroying it mainly because of a failure to recognize networks. Instead, they cling to linear thinking which obscures the full context and consequences of their actions.

(9) To end this much too brief discussion, I would like to present the example of fire-walkers in the context of network thinking. Fire-walkers are people who, in a trance, manage to walk over burning coals without getting a blister on their feet (see, e.g., Grosvernor and Grosvernor 1966). In one case the temperature of the burning coals was measured and found to be $1,328°$ Fahrenheit (see Pearce 1973, p. 107; LeShan 1976, p. 23). In terms of linear causality according to the traditional definition such events appear like miracles because we are used to looking at fire as the cause for burns. Why someone like a fire-walker would not get burned is difficult to understand in terms of traditional causation. However, the notion of networks provides a framework that illuminates this phenomenon. Since, according to network thinking, an event is dependent on the whole system, one has to take into consideration the whole system for the explanation of an event. In the case of fire-walkers, we have to include the walker and his state of consciousness. It is obvious, then, that fire cannot be considered the cause of burns, because the fire is only a part of the system in which burning occurs. What burns is fire in the context of the system which includes the person who gets burned. A change of that system, by changing the state of consciousness of the person concerned, may change the network interactions in such a way that burning no longer occurs.

I am not claiming that the above presentation constitutes an explanation of the phenomenon of fire-walking. However, I think that it provides a framework which allows us to see that such phenomena may be possible in principle. And, furthermore, this framework may furnish a basis for the eventual explanation of such phenomena and other so-called miracles.

To add one more example of this kind I mention the *Kavadi* bearers in South-East Asia. These are devout Hindus, who, also in a trance, bear a frame with many hooks that penetrate their skin. Furthermore, many of them have their tongue and cheeks pierced with metal sticks. Contrary to our expectations, the injuries "cause" neither pain nor bleeding provided the Kavadi bearer has prepared himself sufficiently to enter the required state of consciousness (see, e.g., Babb 1976; Ward 1984). Hence, it is again the whole system including the state of consciousness of the Kavadi bearer that determines whether pain and bleeding occur or do not occur. It is evidently not correct to say that hooks (or the piercing of the skin by hooks) causes the bleeding and pain, although this is our customary way of looking upon such phenomena in terms of linear causality.

### 6.1.7 Network Thinking in Medicine

I shall now briefly apply the preceding general statements to two concrete examples from medicine. In the first example, the problem of cancer, I shall proceed from the most simplistic approaches and interpretations to the more refined and shall show how philosophical assumptions are reflected in the kinds of questions we ask.

A common question that one can hear is: What is (are) the cause(s) of cancer? Although this question may be interpreted differently, it may imply the deeply engrained notion of linear causality. In that case it cannot help us to understand the phenomenon of cancer because the development of malignant tissue entails network interactions.

Approaching the cancer problem in terms of networks will lead to a number of important consequences. We shall have to investigate the whole system and thus use the approach of systems research and systems analysis. We shall have to take into consideration all factors of the system and we should not neglect the environment including the investigator or physician who is trying to heal the patient suffering from cancer. We shall not restrict ourselves to the physical aspect of the system, but shall also consider the influence of mental and psychic factors, i.e., the state of consciousness. Psychosomatic medicine, which is not yet practiced sufficiently, is an important consequence of this more comprehensive approach. It would be misguided to search for one factor such as one substance that could cure cancer in all cases. First of all it is not only one or a few factors that cause cancer or its cure, but the whole system. And secondly, since the system of one patient may differ in some ways from that of another patient, different treatment may be required to achieve the same result. This emphasizes the need of individualistic treatment.

The systems view of cancer also provides a broader perspective with regard to divergent medical practices. Patients have been cured from cancer by changing their state of consciousness due to meditation and other spiritual practices. For example, Oki Sensei, the author of Oki-do Yoga, explained in a Yoga camp which I attended that his intestinal cancer disappeared while he practiced meditation and Oki-do. Since the state of consciousness is an all pervasive component of the system, it is in principle possible that its change may lead to the cure of cancer. Other cancer patients have been cured while following certain diets such as the living foods diet or the macro-biotic diet (see, e.g., Kushi 1978, 1981; Kohler and Kohler 1979). We may ask ourselves whether modern Western medicine will be successful in dealing with the problem of cancer mainly from a biochemical point of view. Mo-lecules of the cancerous tissues are of course also factors in the system, but they are only parts of the whole that should be viewed much more com-prehensively.

The second example concerns impaired vision such as myopia. Usually it is considered to result from a defect of the eyes. Thus doctors tend to deal only with the eyes and may prescribe corrective glasses without actually improving the original organic defect. It has been shown, however, that poor vision may be improved not only by exercising the eyes, but through con-centrated movements of the whole body. Successes of this kind are under-standable from the point of view of systems theory and network thinking. Since eyes and body are part of one integrated system, it is indeed myopic to restrict one's attention only to the eyes of the patient instead of consider-ing the whole person in his or her environment.

It is evident that the notion of network thinking in connection with sys-tems thinking is of paramount importance for modern medicine and has far-reaching consequences. Tremblay (1983, p. 6–7) pinpointed some of the consequences in the following way: "(1) illness can be a global phenomenon; (2) the causal process is diffuse, multifactorial, and socio-cultural ...; and (3) the treatment involves both the modification of life conditions of the afflicted and the modification of the society in which an individual lives" [for further reading on this topic see, e.g., Illich (1976); Engel (1977), Men-delsohn (1979)].

## 6.2 Causalism

Causalism refers to the strict principle of causality which can be stated as follows: "Every event is caused by some other event" (Bunge 1979a, p. XXII). Evidently, causalism presupposes first of all the notion of causation and secondly the postulate that causation is universal, i.e., that there are no gaps in the causal interconnection of events. Both assumptions are questionable. If causation is understood in traditional terms, thus implying linear causality, no further discussion is necessary because, as I have tried to show, this tradi-tional view of causation seems generally untenable. Apparently the principle of causality relies on a traditional notion of causation. It is difficult to imagine how it could refer to network causality in the above definition.

According to Bunge (1979a, p. XXII), the second assumption of causalism is also questionable. Bunge referred to the following two widespread phenomena that curtail the universality of causal interconnections: (1) spontaneity (or self-movement or self-determination), and (2) chance.

In spite of all these restrictions one can still read in many books that the principle of causality is valid, or that, although its validity cannot be demonstrated conclusively, biology as a science is only possible if this principle is assumed to be correct. Mohr (1977, p. 76) also insists that the principle of causality is a sine qua non of science, although he concedes that it is a difficult question to decide whether the principle is true. It is, however, important to note, that Mohr defines the principle of causality quite differently. To him causalism is the same as determinism. Thus, the "formulation of the principle of causality as a negation is: There are no indeterminate events" (Mohr 1977, p. 76). In order to examine whether the principle of causality is tentable in this much wider definition that is compatible with the notion of network causality we have to analyze first the notion of determinism.

## 6.3 Determinism

Determinism can be defined as a doctrine according to which all events including human actions are determined by the totality of preceding events. From this doctrine it follows that if we knew all present events (and the laws that govern them), we could predict all events of the future. Our present inability to predict the future completely, according to this doctrine, results from our insufficient knowledge of the present events. As we learn more and more about the events of the present, our prediction of the future would become increasingly accurate. In classical physics this doctrine was expressed by the symbol of Laplace's demon. This demon would know the present positions and velocities of all particles in the universe and would be a mathematician of unlimited competence. He therefore could calculate, i.e., predict, all events of the future. Thus, the universe was thought to be a deterministic mechanism.

Discoveries in physics during our century have undermined this deterministic picture of the universe. We now think that it is impossible to determine exactly the position and the velocity of a subatomic particle at a certain time. Hence, an exact prediction of the future positions and velocities appears to be principally unattainable. We can make predictions of future states only with certain probabilities, not with certainty. Thus, in microphysics we have to accept the uncertainty principle. This principle, named after Heisenberg, is one of the great discoveries (inventions) of physics in our century. Its philosophical impact has been such that our century has been called the century of uncertainty in contrast to the preceding centuries of classical deterministic physics. It is important, however, to underline that physicists and philosophers of science do not agree on the ontological interpretation of this principle. Two quite different conclusions can be drawn from it with regard to reality:

(1) The principal impossibility to predict accurately future states of subatomic particles is considered to indicate that nature is indeterministic. Thus, indeterminism is a reality and not only the reflection of a methodological limitation. (Indeterminism has also been called acausality; however, it is preferable to refer to indeterminism in order to avoid the general problems of traditional "causality.")

(2) The uncertainty principle only reflects our limited knowledge as far as prediction of future states is concerned. Nature actually is deterministic. This viewpoint was expressed by Einstein's famous statement: "God does not play dice." It is evident that this statement is a matter of faith, which is not well supported by more recent experiments (see, e.g., Davies 1983, Chap. 8). Eigen and Winkler (1981) accept that God plays dice. However, they insist that he obeys the rules (laws) of the game.

In very simple terms the two interpretation of the uncertainty principle can be expressed as follows: (1) complete determination does not exist in reality, (2) it exists, but we have no means of knowing it.

Although Heisenberg's uncertainty principle refers only to microphysics, the same kind of reasoning could be applied to higher levels of organization from cells and organisms to whole societies. And again two interpretations are possible for the uncertainty of prediction. Thus, in biology, psychology, and the social sciences, the failure to make good predictions of the future could be attributed solely to inadequacies and limitations of our present knowledge and methodologies, or it could be argued that the difficulty of predicting future states is also due to an indeterminism inherent in nature (see, e.g., Stent 1978, Rigler 1982).

Instead of concluding that modern science has lead us to recognize a certain indeterminism (which can be interpreted as methodological or ontological indeterminism), we could also state that modern science has forced us to change our notion of determinism (Bunge 1979a). This latter conclusion amounts to a much broader definition of determinism that includes not only strictly determined phenomena (= determinism s. str. discussed so far), but also phenomena that occur with a certain probability such as microphysical events to which Heisenberg's uncertainty principle applies.

If this broad definition of determinism is accepted, then four types of determination can be distinguished whose description by Bunge (1968, p. 311) is as follows (P stands for 'property'):

1(a). Simple action: $P_1$ and $P_2$ determine $P_3$.

1(b). Stochastic action: $P_1$ and $P_2$ determine the probabilities of $P_3$.

2(a). Simple interaction: $P_1$ and $P_2$ determine $P_3$, and conversely $P_3$ and $P_2$ determine $P_1$.

2(b). Stochastic interaction: $P_1$ and $P_2$ determine the probabilities of $P_3$ and conversely, $P_3$ and $P_2$ determine the probabilities of $P_1$.

Since Bunge's classification of the types of determination is very comprehensive, it is most useful to summarize my discussion on causality, causalism, determinism, and indeterminism. "Only *simple action* [case 1 (a)] can qualify for causality: the remaining types of determination are far more complex than causal determination" (Bunge 1968, p. 313). We must conclude then, that "causation has turned out to be a very special kind of determination and, moreover a nexus that is far from being universal" (Bunge 1968, p. 315). It is the kind of determination that in my opinion is of little or no importance in complex living systems. In contrast, simple and stochastic interactions are notions of great importance in biology because these notions refer to the networks of living systems. "Stochastic interaction" probably plays a more important role than "simple interaction" because most of the biological laws appear to be probabilistic laws. Stochastic interactions imply indeterminism s. str. as it is exemplified in Heisenberg's uncertainty principle and other uncertainties at higher levels of organization.

Bunge's characterization of the four types of determination is simplified in considering only three properties, $P_1$, $P_2$, and $P_3$. Naturally, these three properties stand for the whole set of properties of the system to be investigated. This means that if we consider the most inclusive system, which comprises the environment of the living systems with the investigator, the state of consciousness of the investigator also has to be considered in interactions. It is not at all an easy task to conceptualize states of consciousness in terms of properties and one may wonder what extent of reality will get lost in this process of abstraction.

With regard to Bunge's conceptual framework of determination in the wide sense, the relations between the principal notions discussed can be presented as follows:

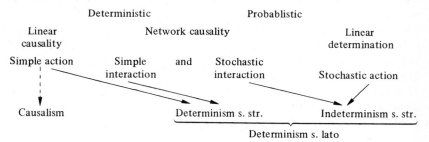

Traditional causality is the prerequisite for causalism, although the latter surpasses the former by the additional assumption of the universality of the causal relation. The principle of causality as defined by Mohr (1977, p. 76: "There are no indeterminate events") could be equated with determinism s. str. or determinism s. lato depending on the implied definition of determinism.

Indeterminism can also be defined in a narrow and a wide sense. So far, I have used the term 'indeterminism' only in its narrow sense which implies that the interactions of things or events are stochastic (probabilistic). Stochastic interactions, although not fully determined in a strict sense, are not chaotic. They can be represented by probabilistic laws, i.e., lawfulness in contrast to chaos which, according to Bunge (1979a), is the total lack of any lawfulness.

Indeterminism in the wide sense refers to chaos, i.e., the lack of lawfulness or the lack of order. As I mentioned already in Chap. 2, it is a highly controversial question whether reality is lawful or devoid of order. Bunge (1968, p. 315) insists that reality must be considered to be lawful even if our direct experience may give us the opposite impression. "No matter how chaotic an arbitrarily chosen set of events may be, the world as a whole is basically ordered" (Bunge 1968, p. 315). Bunge admits that this hypothesis of the lawfulness of the world is metaphysical. It must be taken as an act of faith. Moreover, Bunge considers it to be "irrefutable for, if anything should look lawless or seems to come out of the blue or go into the blue, we would ask for, and be granted, all the time necessary to refute such an impression" (Bunge 1968, p. 315). If we are prepared to stick to this faith we may indeed die with this conviction without ever having seen the magic blue of the sea and the sky.

Being immersed in the magic and mysterious blue of the Indian Ocean and sky while writing this section on determinism and indeterminism, I am indeed wondering whether nature is lawful or chaotic. In any case I see no reason why we have to be dogmatic with regard to this question. Bunge (1968, p. 315) insists that the hypothesis of the lawfulness of nature (i.e., determinism s. lato) is a prerequisite for scientific research. But if in the future instances of lawless phenomena recalcitrant to the deterministic (s. lato) approach should become increasingly frequent, would it not be our responsibility as critical scientists to accept a critical attitude toward this metaphysical hypothesis? Would it be wise to continue doing science based on lawfulness in the face of repeated objective experience to the contrary? Would we not reach a point at some stage of continued disappointment where we would either give up science completely, or acknowledge that it is only of limited usefulness because nature is only lawful to a limited extent? Stent's (1978) contention that scientific progress will soon come to an end is partly based on the lawless aspect of living systems.

However, insisting on the lawlessness and disorder of nature would be just as dogmatic as an insistence on order. I think we have to remain open-minded with regard to this question. Furthermore, we have to keep in mind that law and order are relative notions: what to one person is chaotic, to another one may still be ordered (see Sect. 8.5.2). We should not take our present notion of order as absolute and unchangeable. Future discoveries may change our notion of order considerably. Hence, what is considered as chaos today, may be conceived of as a new form of order tomorrow. The historical development of science supports this contention. In the nineteenth century, order was equated with determinism s. str. In our century we had to adopt a broader view of order. At the time of Linnaeus, taxonomic order was seen in terms of the fixity of species. Today's taxonomic systems are a far cry from the narrowly conceived order of Linnaeus' time (see Sect. 4.2.1). The order of logic also has undergone profound change from the simple bimodal logic in terms of true and false to multivalued logic and fuzzy-set theory.

I think that much of the controversy over the presence or absence of order in nature is a reflection of different ideas of order and chaos. Hence, much of the disagreement is probably semantic. However, apart from that, it will remain a perpetual fundamental problem for scientists whether nature exhibits some kind of orderliness. After all it is the aim of science to extract regularities of some sort from the welter of experience. In this attempt scientists have to listen to nature as much and as openly as possible. If they succeed in transcending their biased preconceptions, they may indeed advance our understanding of nature beyond our imagination, just as our contemporary science is beyond the imagination of the nineteenth century scientists. As Bunge (1968, p. 315) pointed out, we used to equate (traditional) causality with lawfulness and with determinism. Both equations had to be given up, because (traditional) causality turned out to be only a special case of lawfulness and determination. In other words: we had to broaden our notion of causality to include network causality, at which point the original term may be dropped altogether. Moreover, we equated causalism with determinism. This was again an equation which had to be given up because causalism turned out to be only a special case of determinism s. lato.

Bunge pointed out that notions like the principle of causality, which today we recognize as being of limited scope, may still be used as a heuristic principle, i.e., as a working hypothesis. "But every time we adopt it as a working hypothesis and as holding to a first approximation, we find something – often a noncausal relation satisfying a richer type of determination" (Bunge 1968, p. 315). The same could be said with regard to lawfulness and order. Provided we remain open-minded and adaptable, our search for lawfulness and order as we understand it now may lead us to the invention of relations satisfying a richer and broader type of lawfulness and order that may come increasingly closer to what today we label as chaos in a more or less derogatory way. To achieve this we have to respect nature whether she appears lawful or chaotic.

## 6.4 Free Will

Am I free to write this section on free will or is it an illusion that I think I made this decision? Generally speaking: is free will real or illusory? This question has been much debated and as the discussion continues there are still those who believe that free will exists and those who deny its reality (see, e.g., Berofsky 1966; Settle 1973; Davies 1983; Lewontin et al. 1984, p. 287). Many of those who claim that free will is illusory refer to determinism as a basis for their position. If everything is determined by prior events, they say, free will cannot exist because it would constitute a break in the determination of events.

Those who defend the existence of free will do it for various reasons. I shall present only two kinds of argumentation, a scientific and a philosophical one. The scientific argument refers to indeterminism s. str., i.e., stochastic actions and interactions. This indeterminism at the microphysical level is interpreted as a reality. Furthermore, it is postulated that this indeterminism is amplified to higher levels of organization and that our willed behavior represents the amplification of "freedom" at the microphysical level (see, e.g., Jordan 1941, 1972). This amplifier theory of free will is rather speculative because the relevant amplification has not been demonstrated scientifically. Schrödinger (1967, p. 92), for example, stated that "quantum indeterminacy plays no biologically relevant role in them [i.e., the space-time events in the body of a living being corresponding to the activity of its mind], except perhaps by enhancing their purely accidental character in such events as meiosis, natural, and X-ray induced mutations and so on." Moreover, it is hard to see how chance events at the microphysical level can constitute a basis for free will that does not appear to be a matter of chance, but as the term indicates, "willed."

The philosophical argument for the existence of free will that I shall briefly present is that of existentialist philosophers. To them the fundamental experience of our existence has its roots at a very basic level at which we have not yet divided the world into subject and object. Free will is experienced in this totality. Science, however, according to Jaspers, for example, always operates within the subject-object division. This means that all statements of science including determinism are statements of subjects about objects. As such they cannot refer to a level of reality that precedes the subject-object division of scientific methodology. This means that scien-

146

tific statements in principle cannot refer to the basic experience of free will. Science is irrelevant as far as the problem of free will is concerned. Our existential experience reveals to us the potential of free action. It is up to us whether we seize upon this potential and live a free life.

## 6.5 Beyond Free Will

Quite often we take it for granted that *either-or* thinking is appropriate. Accordingly, we assume that free will must either exist or not. The only task, then, is to decide which of the alternatives should be accepted. It is, however, also possible that the alternative does not make sense or does not represent a good or useful question. After all questions are not given, but are dependent on logic, conceptualizations, and other presuppositions. Inadequacies in these foundations may lead to questions that are meaningless (i.e., pseudo-questions) or a distorted and misguided inquiry.

I shall now present the point of view of Schrödinger (1967, Epilogue) on the issue of determinism and free will. Normally these two phenomena are seen as mutually exclusive (unless one takes the point of view of existentialists who refer to two totally different levels). Schrödinger's insight reveals how these phenomena can become noncontradictory. He starts out with the following two premises:

(1) "My body functions as a pure mechanism according to the Laws of Nature."
(2) "Yet I know, by incontrovertible direct experience, that I am directing its motions, of which I foresee the effects, that may be fateful and all important, in which case I feel and take full responsibility for them."

The solution of the apparent problem presented by these two premises that appear to be contradictory is the following according to Schrödinger: "I ... am the person, if any, who controls the 'motions of the atoms' according to the Laws of Nature." This means that the dichotomy between myself and the universe is annihilated or has become transparent, i.e., I realize that it has never existed but is the result of conceptualization. In terms of Indian philosophy of Vedanta this is "the recognition of ATHMAN = BRAHMAN (the personal self equals the omnipresent, all-comprehending eternal self)" (Schrödinger 1967, p. 93). This unity of the personal self, which experiences free will, and the eternal self, which represents the Laws of Nature, annihilates the contradiction of the two above premises.

Schrödinger's insight does not amount to a rational solution of the problem. However, a problem as basic as this cannot be expected to be solvable just by rational means. It reaches too deeply into our subconscious and the

core of reality. Mohr (1977, p. 90), in discussing the contradiction of Schrödinger's premises, i.e., determination and free will, also admits that "the
problem free will vs determination is a problem that cannot be solved by
rational argument, by thought." Schrödinger points out that "the mystics
of many centuries, independently, yet in perfect harmony with each other
... have described each of them, the unique experience of his or her life in
terms that can be condensed in the phrase: DEUS FACTUS SUM (I have become God)." Being God myself, I can no longer be controlled by an external
God and/or (his) Laws of Nature.

One might interpret Schrödinger's conclusion as a justification for the
existence of free will. However, I think that this would be a mistake. Free
will is a phenomenon of the experience of the personal self. As the dichotomy between personal self and eternal self is transcended in the awareness
of the One, the dichotomy between free will and determination vanishes; it
is seen as a conceptualization instead of two separate phenomena of reality.

"Yoga [i.e., union of personal and eternal self] is thus an extraordinary
state of being ... It is extraordinary only in contrast to the ordinary conditioned state of being in which man is born and brought up. This conditioned
state is maintained by the pressure of the prestigious norms of social con-
formity, but it can be challenged by the man who is willing to do so in the
interest of discovering the truth which underlies his conditioned and conformist way of living" [Deshpande (1978), p. 4 on Patanjali's Yoga Sutras].

## 6.6 Summary

*Causality.* Causality continues to be a controversial issue among both philosophers and scientists. Whereas some consider it a "relic of a bygone age,"
others see a revival of causal thinking, especially in the life sciences. Regardless of which point of view one takes, one can hardly deny that the notion
of causality has undergone profound changes so that it is indeed debatable
whether the modern notion of network causality is still related to the traditional notion of causality or whether it supersedes it completely which I
think is the case. Traditional causality (in the sense of Aristotle's efficient
cause) refers to the relation of cause and effect in such a way that (1) the effect follows the cause in time, (2) the cause produces the effect, and (3) the
relation between the two is constant and necessary. The interconnection of
causes and effects produces causal chains. This view of causal interconnections constitutes the conventional notion which is referred to as linear causality. Apart from branched chains, more recent research has necessitated the
use of the notions of circular and network causality. These notions differ
fundamentally from the traditional concept of linear causality so that in the

148

case of network causality it is better in my opinion to refer to network interactions and network thinking. One corollary of network thinking is that any
particular thing or event is determined by the whole network, i.e., the whole
system. In terms of traditional causality this means that everything (i.e., the
whole) is the cause of a particular thing or event. Obviously, such a statement stretches the notion of traditional causation to a point near meaning-
lessness. One has to consider, of course, that certain segments of the network may have a weak (hence negligible) influence on a particular change.
However, the whole must be kept in mind when we are dealing with parts.
According to Bohm (1980) the diversified and fragmented order of things
and events we observe (i.e., the explicate order) is the unfoldment of an
underlying implicate order which is holomovement, i.e., "undivided wholeness in flowing movement" (Bohm 1980, p. 11). This idea goes even beyond
network thinking because the points and lines in the net merge (enfold) into
undivided wholeness.

Mohr's model of causal analysis, which is actually a factor analysis, appears to be quite compatible with network thinking. The consequences of
network thinking are profound and manifold. It emphasizes the need for
systems thinking which implies structural and functional integration of parts
and subsystems into more and more inclusive systems that eventually have
to include the environment of organisms with the investigator who is also an
integral part of the whole system. The methodological, scientific, and philosophical consequences of this are mind-boggling. Existential and social consequences amount to basic transformations of human beings and society that
require a more holistic view of human behavior and social action. Some of
the far-reaching consequences of network thinking are exemplified with
reference to cancer and impaired vision. The importance of an integrated
approach to healing is pointed out. This integration must include body and
mind-soul as well as patient and environment which comprises the physician
and investigator.

*Causalism.* Causalism refers to the strict principle of causality which can be
stated in the following form: "Every event is caused by some other event"
(Bunge 1979a). In this form the principle is untenable because it implies the
notion of linear causality and its universality. The latter can be questioned
because of the phenomena of spontaneity and chance. Mohr (1977) defines
the principle of causality in a wider sense in terms of determinism as follows:
"There are no indeterminate events."

*Determinism.* Bunge (1968) distinguishes the following four types of determination: simple and stochastic action, simple and stochastic interaction.
Linear causality and the strict principle of causality (sensu Bunge) are based

on simple action. Determinism s. str. implies simple action and simple interaction. Indeterminism s. str. (to which Heisenberg's uncertainty principle refers) is characterized by stochastic action and stochastic interaction. Determinism (sensu Bunge) implies all four types of determination, i.e., it comprises determinism s. str. as well as indeterminism s. str. Indeterminism s. str. can be interpreted methodologically or ontologically. In both cases it entails order which, however, is no longer strictly deterministic if the indeterminism is interpreted ontologically, i.e., residing in nature. Indeterminism s. lat. amounts to lack of order, i.e., chaos. Several questions of fundamental importance arise with regard to the problem of order and chaos. Is nature ordered or chaotic? Is order in nature an absolute prerequisite for scientific research? What are our notions of order and chaos? If these notions are relative and continue to change, phenomena that today appear chaotic may reveal aspects of order (in a new sense) in the future.

*Free will.* Free will also continues to be a controversial issue. Many of those who deny its existence interpret determinism ontologically and believe that there are no gaps in the determination of events. Those who defend the existence of free will may do so for scientific or philosophical reasons. One scientific argument traces free will to indeterminacy at the microphysical level which is thought to be amplified to the macro level and thus is supposed to constitute the phenomenon of free will. Existentialist philosophers such as Jaspers point out that free will is an existential experience at a level more basic than the subject-object division which is a prerequisite for scientific research. Hence, free will cannot be the subject of science.

*Beyond free will.* According to the physicist Erwin Schrödinger, the contradiction between determination and the experience of free is only an apparent one that is resolved through mystical union of the individual person with the universe and its laws. Hence, it is I "who controls the 'motions of the atoms' according to the Laws of Nature." Such has been the experience of the mystics of millenia.

# 7 Teleology

> "But the only real danger to be feared lies in being too easily satisfied with the belief that the last [final] word has been said on this topic" (Woodger 1967, p. 451)
>
> "Seeking means: to have a goal; but finding means: to be free, to be receptive, to have no goal. You, O worthy one, are perhaps indeed a seeker, for in striving towards your goal, you do not see many things that are under your nose" (Hermann Hesse: Siddhartha, 1957, p. 113)

## 7.1 Introduction

Besides the question "what is the cause?", one of the most common questions asked by students of biology is: what is the purpose (function)? It is often assumed that if we just knew the purpose or function of a structure or process we would have really understood it.

Teleology refers to purposes, functions, and related issues which in turn refer to the most complex phenomena of living systems. It is therefore not surprising that, as Ruse (1982) put it, teleology is a "hot" topic which, according to Stegmüller (1969, Vol. 1, Part 4, p. 518), is a nearly unpenetrable jungle. The literature on teleology is immense. Some of the more recent discussions are by Canfield (1966), Beckner (1969), Monod (1970), Grene (1974), Wright (1976), Woodfield (1976), Mayr (1976a), Bernier and Pirlot (1977), Nagel (1979), and Brandon (1981) (for further references see Grene and Burian 1983). The following two aspects may be distinguished in the multitude of controversies on this subject matter:

(1) Disagreement concerning the terms and concepts employed. A number of terms that refer to a host of more or less different concepts have been used. Thus the semantics of teleology has become nearly chaotic.

(2) Disagreement concerning the interpretation of the phenomena to which the multitude of teleological concepts may refer directly or indirectly. Among these phenomena are homeostasis, homeorhesis, self-regulation, adaptation, survival, selection, integration, organization, goal-directedness, feedback, and programs. Hull (1974, p. 103), in an attempt to establish properties characteristic of all teleological systems, enumerated them in terms of the following four criteria: "prevalence of preferred states, closed feedback loops, and programs, and the origin of such systems by means of selection processes." Being aware of the semantic difficulties alluded to already, he added that "we cannot set out a list of criteria which are severally necessary

and jointly sufficient for a system's being teleological. To no one's surprise, I am sure, we have not discovered the essence of teleological systems" (Hull 1974, p. 103). This in turn is not surprising to Bernier and Pirlot (1977) because according to them there is nothing like the essence of teleology in nature except in organisms such as human beings that are endowed with intentionality. They accept Hull's four criteria (properties) as being characteristic of organic systems (Bernier and Pirlot 1977, p. 73), but they deny that these four properties indicate anything in living organisms (lacking intentionality) that is teleological in any sense beyond the four properties. Therefore, they see no reason whatsoever to use the term teleology or related terms because such terminology would tend to create confusion inasmuch as it might indicate that there is a phenomenon that according to them is purely fictitious. Not all biologists and biophilosophers agree with this contention of Bernier and Pirlot. Thus, the debate over teleology (regardless by which name it is called) concerns also an aspect that is more than just pure semantics. It is a debate over the nature of living systems. However, before discussing these interesting questions, I shall attempt to clarify to at least some extent the semantic confusion surrounding the notion of teleology.

## 7.2 Terminology

As far as the diversity of definitions and meanings of teleology and related terms is concerned, one wonders whether anybody today is still able to cope with the semantic confusion that has resulted from it. Etymologically, the word teleology is derived from the following two Greek words:

telos = end, goal
logos = reason, doctrine

Thus, the literal meaning of teleology is the logos of ends. At this point problems arise already because 'end' can be defined differently. Accordingly, different concepts of teleology are established and consequently different approaches to the phenomena of life may be adopted.

If we observe a sequence of states (or events) that begins with state A and proceeds via states B and C to state D, i.e., $A \rightarrow B \rightarrow C \rightarrow D$, we may call state D the end of this sequence. If we make the repeated observation of this sequence, we may make the inductive generalization that in general D is the end of such a sequence. If, in addition, we observe that D is even reached when instead of B or C, B' or C' or other stages are passed, then we might become convinced that D is not just the end of a sequence, but that the events are directed towards D as their goal. In this sense the adult stage of living organisms may be looked upon as the final stage, i.e., the goal of their development (ontogeny). End-directed (or goal-directed) behavior could be understood in this way. This may, however, create confusion since the term 'directed' can be interpreted in quite different ways. Accordingly, different concepts of teleology can be distinguished:

(1) Laws of nature are held responsible for the goal-direction of the sequential behavior which is also called developmental teleology (Grene 1974, p. 211). The role of consciousness is excluded. For example, the regular recurrence of the goals (i.e., "preferred states" according to Hull 1974) is seen as the result of feed-back mechanisms and the natural selection of those mechanisms (see, e.g., Monod 1970; Bernier and Pirlot 1977). Genetic programs are also considered to be important in this respect because they are selected (see, e.g., Mayr 1976a; Bernier and Pirlot 1977). If this kind of definition of teleology is adopted, it might be affected by a change of our biological theories, especially morphogenetic and evolutionary theories.

(2) Intentionality, which presupposes consciousness, is held responsible for goal-direction. For example, I reach the goal of writing this book because I consciously anticipate my goal and then work toward it. This kind of teleology is generally accepted for humankind (also by Bernier and Pirlot 1977), but it is debated whether it applies to animals and plants (see below). In order to avoid semantic confusion, I shall refer to this second concept as goal-intendedness (see Stegmüller 1969, Vol. 1, Part 4; Wuketits 1978, p. 128). The term 'goal-directedness' I shall use only in the sense of definition no. 1. If intentionality presupposing consciousness is considered a law of nature, the teleology based on it becomes a special case of the first category of definition.

(3) A vital force immanent in the living organism has been held responsible for goal-direction according to vitalistic thinkers (see, e.g., Driesch 1908).

(4) Finally, the direction toward goals is seen as the act of a god, creator, or mind who determines the goals either directly through constant interference or indirectly by the creation of systems so designed by him that their continued teleological behavior is guaranteed without his further interference.

Most modern biologists reject the third and fourth concepts of teleology, and as far as plants and (the majority of) animals are concerned they are also unwilling to apply the second definition (see below). Hence, in biology mainly the first definition is used. In order to distinguish it from the others, the term 'teleonomy' has been coined by Pittendrigh (1958). Mayr (1974, 1976a) defines teleonomy still more narrowly as referring only to those goal-directed processes that are initiated and controlled by a genetic program. Broader definitions of teleonomy would imply a broader view of development and evolution (see, e.g., Waddington 1970, 1975; Wuketits 1978). Bernier and Pirlot (1977) insist that even the term 'teleonomy' is unnecessary and furthermore confusing because it implies "goal-directedness" which to many people has the connotation of intentionality. However, during the last decades a fairly large number of biologists and philosophers have been using the term 'teleonomy' in a sense that does neither imply intentionality, nor vital force, nor external supernatural forces (see, e.g., Ayala 1968; Mayr 1969, 1976a, 1982; Monod 1970; Jacob 1970; Ruse 1973; Hull 1974; Wuketits 1978; Riedl 1980).

Woodger (1967, p. 436) distinguished internal and external teleology (see also Ayala 1968). Internal teleology resides within the system under consideration, whereas external teleology results from a system that is external to it. For example, certain sophisticated machines (servo-mechanisms which can be called human artefacts) such as goal-tracking missiles, exhibit a striking teleological behavior in terms of goal-directedness; this teleology originates from the intention of the men who built the missiles. Thus human intentionality and action created the missiles, and in this sense the teleology (i.e., goal-directedness) of missiles is external. According to the fourth definition above, the teleology of organic nature has been understood in a similar way as having its origin in a creator who designed the teleological systems.

With regard to human artefacts, I think it might be useful to distinguish two aspects. From the point of view of human intentionality it is a case of external teleology. However, from the point of view of teleonomy (i.e., our first definition) we may look upon it as internal since the functioning of the missiles is independent of our intention; it depends on the cybernetic system that we have constructed for this purpose; our intentional interference is no longer required except that we push certain buttons.

In the above discussion of goal-directedness according to laws of nature, regular sequences of events with a common end-state (goal) have been considered. In addition to regular sequences, however, we have to take into consideration unique sequences. These events may also be important for the survival of organisms (Woodger 1967, p. 439). If we say now that a unique event "just happened and was accidentally appropriate, this is equivalent to saying that *no* answer *can* be given. And it may be that this is all that it will ever be possible to say ... because human thought (perhaps) cannot deal with particular (i.e., unique) occurrences" (Woodger 1967, p. 439).

Teleology has not only been defined in terms of goals, but also in terms of purpose. In this sense teleological behavior is purposive behavior. Teleology thus refers to purposiveness. "Purpose," in turn has been equated with "function." However, it is also meant to be "that which one sets before oneself as a thing to be done or attained" (Oxford Dictionary), or "something set up as an object or end to be attained" (Webster's Dictionary). Especially the latter definition makes it evident that we encounter here the same kind of problems as with regard to goals, namely the question of which agent set up the goals. Depending on the answer, purposiveness implies goal-intended-ness (due to intentionality), or a vital force (e.g., Driesch 1908), or the inter-vention of an external spiritual force. Thus some of the definitions of tele-ology in terms of purpose may coincide with some of the previous defini-tions of teleology. However, other definitions of purposiveness such as, for example, the equation of purpose with function, deviate from all the previous definitions of teleology. Grene (1974, p. 211) refers to instrumental tele-ology in this context.

In defining teleological behavior as functional behavior, the main ques-tion is the definition of function. The following definitions may be listed. Needless to say, this list is incomplete.

(1) As pointed out already, function may be used as a synonym of 'pur-pose.'

(2) Function is defined as a property of a structure. Thus, according to Bernier and Pirlot (1977, p. 38), the function of an organ is a property of that organ.

(3) According to Beckner (1969, p. 151) who uses a functional defini-tion of 'teleology,' 'function' is "the role of a part or process in the activities of a more inclusive system."

(4) The preceding definition implies often (but not necessarily) that the role a part or process plays in the whole system is important for the economy or survival of the whole (see e.g., Woodger 1967, p. 327; Zimmermann 1968, p. 72; Wuketits 1978, p. 127; Riedl 1980, p. 152). In that sense the function is defined with regard to adaptation and survival. However, according to Bernier and Pirlot (1977, p. 41), the role that a part or process plays does not define function, but the (biological) role is an ecological attribute that a function may or may not have. For example, vision is a function that can play a different role in the ecology of different kinds of organisms. Further-more, not all functions play a role, i.e., are important for survival.

(5) According to Woodger (1967, p. 327) who distinguished three defini-tions of 'function,' 'function' can also stand "as a name for all the processes ordinarily said to be 'going on in an organ." For example, "this is what physiologists usually study when they are said to be studying the physiology of the kidney" (Woodger 1967, p. 328). Thus, function in this sense simply means "activity" in a living system.

(6) Finally, this definition by Woodger could be formulated in such a general way that any process or action within an inclusive integrated system is considered a function. In this sense "function" is not restricted to living systems. [By the way, processes concerning nonliving objects that reach a final point as the result of the operation of a natural law are called tele-omatic processes (Mayr 1974). An example is the fall of a stone.]

Function is often seen as an antithesis to structure (see Woodger 1967, p. 326). However, this antithesis is not real as can be shown in different ways. According to Bernier and Pirlot (1977, p. 38), the function of an organ is just one of the many properties of that structure. Since all of these properties form an integrated whole, function and structure are just aspects of the same thing. According to Woodger (1967, p. 328), structure is the result of selective attention to the visual experience of an object at a specific time. Thus structure results from the fixation of the visual experience. It is a visual time-slice of the dynamic living system. For example, the structure of the heart is relative to the time at which it is fixed. If as developmental morphologists we try to synthesize the time-slices of the heart that we have studied, we are still limited to visual experiences of the heart. To overcome this limitation it is necessary to enlarge our concept of structure "so as to include and recognize that in the living organism it is not merely a question of spatial structure with an 'activity' as something over against it, but that the concrete organism is a spatio-*temporal* structure and that this spatio-temporal structure *is* the activity itself" (Woodger 1967, p. 330). If we perceive the heart in this unusually wide context where structure coincides with activity, then "the heart is an event (or a part-event of the event of greater spatial extent which is the organism)" (Woodger 1967, p. 328).

In addition to its definition in terms of goals, purpose or function, teleology (and teleonomy) have also been defined in terms of adaptation. Thus, teleological systems are adapted systems (see, e.g., Mayr 1974, 1976a; for a criticism of this definition see Bernier and Pirlot 1977, p. 100–102). Brandon (1981) recognizes explanations of adaptations as teleological explanations.

Another term that is often used as a synonym to various concepts of teleology or may be used in more or less different ways is 'finality' (see Gilson 1971). Jacob (1970), to whom finality is a characteristic feature of living systems, defines it in terms of projects that are determined by the genetic program. No intentionality is involved here. However, some other authors believe in various agencies, such as intentions, vital forces and spiritual energies as determinants of finality. Aristotle defined finality in terms of final causes. Riedl (1980, p. 165, 183) also refers to final causes with regard to purposiveness in the sense of functionality.

In conclusion, we may cite Hull (1974, p. 103) who wrote that teleology refers to "all statements couched in terms of goals, purposes, and functions," whereby these terms may be defined differently as indicated above. Accordingly, teleological statements may be formulated using phrases such as "the function of," "the purpose of," "for the sake of," "in order that," "so that," "made for," "designed for," "the goal of," "the role of," etc. For example, teleological statements concerning the eye may be: "the function of the eye is vision," or "the eye exists for the sake of seeing," or "we have eyes in order that we can see," or "the eye is made for seeing," etc. Unless the author who uses such phrases defines their meaning and/or explicates the implied notion of teleology (or teleonomy), they are ambiguous and create confusion because two identical statements may have quite different meanings, whereas statements couched in different terms may be synonymous. As long as the terms are well defined I find no ambiguity even in a usage of terms that may be different from the traditional meaning. We know that the meaning of terms may also evolve and change. There are many examples in science that illustrate how terms have acquired differing meanings. For example, in physics the term 'force' is generally used in a specific sense that differs from the traditional usage of the term with a variety of connotations that may also entail the magical and supernatural. Similarly, the term 'teleology' can be stripped of older connotations and can be defined in relatively precise ways as is evident in the definitions of Nagel (1961), Canfield (1966), Ayala (1968), Beckner (1969), Stegmüller (1969), Monod (1970), Mayr (1974), Wuketits (1978), and others.

Bernier and Pirlot (1977) insist that all terms referring to teleology in the widest sense (including the term goal-directedness) should be eliminated because teleological concepts do not apply to plants and the majority of animals. Although I can see that one might wish to introduce another terminology for the sake of clarity, I am not convinced that the various notions of teleology (including teleonomy) are totally inadequate with regard to plants and animals. I tend to think that teleological notions may have at least limited applicability inasmuch as they refer to an aspect of living systems that may not be fully describable in nonteleological language. Besides that, teleological notions may be of heuristic value not only in biology but even in physics (Woodger 1967, p. 432). For example, Whitehead (1925, p. 77) reported that Maupertuis discovered in physics the famous theorem of least action by starting out "with the idea that the whole path of a material particle between any limits of time must achieve some perfection worthy of the providence of God." Whitehead added that "Maupertuis' success in this particular case shows that almost any idea which jogs you out of your current abstractions may be better than nothing" (quoted by Woodger 1967, p. 432).

These considerations on the applicability of teleological concepts and their heuristic value surpass the terminological aspect and reach the heart of the teleological problem, namely the question of whether there is purpose in nature and, if so, in what sense. I shall examine crucial aspects of this problem in the following sections.

158

## 7.3 Goal-intendedness

Human behavior appears to be determined or influenced to a considerable extent by goals which are set and then pursued with the hope of reaching them. The question that arises is how the goals are set as far as the human species is concerned. Mohr (1977, pp. 86–94) distinguished the following elements required for goal-setting: consciousness, rationality, free will, values, propensity structure, and experience. Values and propensity structure are strongly interrelated and propensity is "part of our genetic inheritance" (Mohr 1977, p. 88). Values such as altruism, friendship, and respect for elders are explained scientifically by the highly controversial science of sociobiology (see Chap. 8.7). "However, it could be that there are at least some values that cannot be explained in full by any scientific theory" because they "might be irreducible spiritual values" (Mohr 1977, p. 87). Experience as an element of goal-setting refers to "adaptive behavioral learning during ontogeny, including cultural tradition, and social imitation" (Mohr 1977, p. 88).

I shall refrain from criticizing this list of goal-setting elements at this point and shall proceed to illustrate Mohr's model of human teleology (see Fig. 7.1). After the goal has been set, the next step is "the conscious selection of the means that seem to be required to reach the goal. Genuine (scientific) knowledge and technology derived from it are the most important means at present, even in our personal lives" (Mohr 1977, p. 89). The action sequence that follows "is as a rule not straightforward. Continuous adjustments are usually required if the informational feedback tells the acting person or group of persons that deviations from the anticipated action sequence have occurred. Moreover, the realization of a goal will *inevitably* lead to undesirable consequences, called side-effects. Since these side-effects may grow into massive regressive phenomena, they must be thoroughly considered and compensated for as early as possible after they become detectable" (Mohr 1977, p. 89). However, since man is "not omniscient, ... we must be prepared to take a risk, even the risk of self-extermination. The feedbacks indicated in the model [see Fig. 7.1] will as a rule help us to minimize the risk, but it can never be completely excluded that gross errors in the estimate of future side-effects and dangers will occur, in spite of concerned, responsible, competent anticipation, and consideration of undesirable regressive phenomena" (Mohr 1977, p. 89/90).

I shall now very briefly examine Mohr's list of goal-setting elements because they are crucial for an understanding and criticism of human teleology. Consciousness is important, but one should also keep in mind that to a very great extent our behavior is influenced by our subconscious (see, e.g., S. Freud and C.G. Jung). Thus, not all action that appears to be goal-intended may be

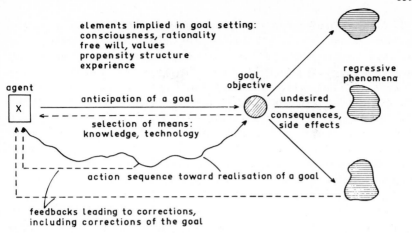

**Fig. 7.1.** Model illustrating the structure of human teleology (goal-intendedness) [After Mohr (1977, p. 87)]

the result of conscious intentionality. Furthermore, it is possible that much of our teleological action is only partially guided by our consciousness. Hence, we have to take into consideration that at least some of the human behaviour may be oriented toward goals without or with only partial awareness of the goals. The following example may illustrate the situation. We may consider it teleological action to study at a university in order to obtain a Ph.D. and to pursue a career as a scientist. Our action is determined or influenced by the goal that we have set for us. We are quite conscious of this goal. However, it is only an aspect of the goal of which we are conscious, namely, the objective aspect of obtaining a Ph.D. If we continue asking ourselves what the goal of a Ph.D. really entails, i.e., why this goal is so desirable to us, then we come soon to the limit of our consciousness. Beyond this limit subconscious urges may play an important role. For example, the myth of the university as described by Novak (1971) may be at the root of our behavior. From this point of view teleological action can be appreciated as a complex behavior and one may ask whether the distinction between conscious and subconscious components is always adequate to understand our actions.

As far as rationality is concerned, similar comments may be made. Much of our behavior has deeply entrenched irrational roots. The question of free will has been briefly discussed already (see Sect. 6.4). Mohr, who doubts the existence of free will, although he emphasizes that "for the sake of our self-esteem we must believe in freedom and free will," concluded that "if we accept the conviction that free will is fiction, it is doubtful whether 'in reality' *we* are able to act teleologically" (Mohr 1977, p. 94).

To what extent our values and our propensity structure are genetically determined has been much debated especially since the appearance of Wilson's "Sociobiology" (1975). It continues to be a highly controversial subject (see, e.g., Caplan 1978; Ruse 1979; Montagu 1980). The influence of experience and hence cultural conditioning cannot be denied, although the extent of it is debatable. Much of what Lorenz (1963, 1973, 1978), Wilson (1975, 1978), and Lumsden and Wilson (1981, 1983) take as inherited values and propensity structures is considered to be due to cultural conditioning by others (see, e.g., Montagu 1980). For example, aggression, which according to Lorenz (1963) is instinctive and therefore inherited, may be largely the result of cultural conditioning (Montagu 1980).

We may now ask: is the teleology as defined by Mohr (1977) and illustrated in Fig. 7.1 applicable to plants and/or animals? I think that most, if not all, biologists would tend to answer this question negatively. As a next step, we may then consider to relax Mohr's definition of teleology by requiring fewer of the elements and/or weakening the elements. For example, we may thus arrive at a definition of teleology that requires for the goal-setting only consciousness, a propensity structure and experience. Since in most cases animals have a propensity structure and learn from experience, the crucial question then is that of consciousness in animals. No agreement exists among biologists or philosophers with regard to this question (Griffin 1982, 1984; Crook 1983; Ferry 1984). However, there are biologists who ascribe consciousness to vertebrates and the more highly evolved invertebrates such as advanced mollusks and arthropods (see, e.g., Griffin 1982). If we accept this belief, then we have to conclude that teleology in the sense of conscious intentionality may exist in a large number of animal species.

We may go further now in our relaxation of Mohr's definition of teleology by treating consciousness as a comparative or quantitative concept so that consciousness becomes a matter of degree. In that case, which I discussed already for human teleological action, goals are conscious only to a certain degree. If they are conscious 100%, we obtain the second definition of teleology (see Sect. 7.2). If they are conscious 0%, then the second definition does not apply at all. However, if the consciousness ranges between 0% and 100%, then the second definition of teleology applies partially. In terms of the notion of a fuzzy set this means that the behavior of certain animals may be only partially a member of the fuzzy set of teleology, i.e., it is only partially teleological. From this vantage point it appears quite inadequate to ask in such cases whether the behavior is essentially teleological or not. This question would be a pseudo-question in this context.

It may be immediately objected that consciousness is an all-or-nothing phenomenon which either exists or does not. If this were correct, then, of course, teleology as a fuzzy set would be an inappropriate concept. However, a number of biologists have speculated that consciousness has evolved gradually and therefore did not emerge suddenly during the evolution of the human species. Inasmuch as this speculation is justified, the above considerations of fuzzy teleology are appropriate.

The next problem is whether teleology in the above relaxed sense of goal-intendedness applies also to plants. This is an extremely difficult problem because it is difficult, if not impossible, to approach it scientifically. However, a number of experiments have been carried out in order to find out whether consciousness and emotions may be ascribed to plants (see Tomkins and Bird 1973). One famous experiment of this sort was the shrimp-killing experiment performed by Backster (1968). Backster dropped living shrimps into boiling water in the presence of a Philodendron plant that was hooked up to a galvanometer so that changes in its electrical potential could be measured. The result of this experiment was rather striking. Each time a shrimp was killed, the plant responded by a change of its electrical potential. Backster performed in addition other experiments by which he tried to demonstrate that plants also respond to human emotions. On the basis of these experiments he concluded that plants must have emotions, and others have become convinced that plants also have consciousness and even may be able to think (see Tomkins and Bird 1973).

In order to evaluate these conclusions it is important to distinguish two components:

(1) the question of whether plants respond to the killing of shrimps, or, as it was the case in other experiments, to human emotions, and
(2) if the first question is answered in the affirmative, the interpretation of this response.

Many biologists have ridiculed Backster's experiments and have implied that it is unscientific to even approach the problem of whether plants may respond to our emotions or the killing of animals. Finally, a team of botanists at Cornell University repeated Backster's shrimp killing experiment with a much more sophisticated experimental design than that of Backster, and their results turned out to be negative, i.e., they could not confirm Backster's findings (Horowitz et al. 1975). Many biologists received this "refutation" of Backster's results triumphantly and considered it the last word on the matter (see, e.g., Galston and Slayman 1979). It is interesting to note that exactly those biologists who accused Backster of the use of an unscientific method tend to consider the results of the team at Cornell University as absolute proof that Backster was wrong and consequently as the final word on the subject matter, as if there could be proof or disproof or anything final in science. Such dogmatism in the name of science is not warranted. Nobody can know what results further repetitions of Backster's experiments would give. One very important point that is overlooked by many of Backster's critics is the attitude of the experimenter(s). Especially in an experiment that involves the problem of the plant's response to human emotions, the emotions of the experimenter(s) cannot be neglected. They might in fact constitute one of the most important factors. Hence, with regard to the first component in Backster's conclusions we still have to wait for more sophisticated experiments. The problem is still open to investigation. I see no reason why it cannot be approached scientifically. Psychology as a science also deals with emotions and response to emotions. What matters is whether our hypotheses are testable, not whether they are about physical properties or emotions.

With regard to the second component of Backster's conclusions, the situation is, however, different. I think that here the methodology was most probably unscientific. Even if it could be confirmed that plants respond to our emotions, it would not follow that therefore plants must also have emotions. One would have to show that a response in terms of the change of the electrical potential of a plant entails emotions of that plant. The same applies to the question of consciousness in plants. A scientific approach to these question appears to be difficult, if not impossible; it is beyond my imagination (which of course does not at all mean that it is impossible in principle). Thus, the problem of emotions, consciousness, or precursors of such phenomena remains a matter of intuition until someone shows how it could be approached scientifically.

I think at this point I have exhausted even the most relaxed version of Mohr's original definition of teleology. Thus, I have to turn now to a definition of teleology that does not imply consciousness (i.e., teleonomy).

## 7.4 Goal-directedness

It is often said that goal-directedness (definition 1, Sect. 7.2) results from the genetic program (see, e.g., Mayr 1974). However, it is also recognized that organic processes are very complex. Thus, the development of an organism is not only directed by the genetic program (i.e., the genome), but is also dependent on the cytoplasm that surrounds the genome and is furthermore influenced by the environment in which the organism develops. Hence, the genetic program is an open "program" (i.e., no program at all): the development of the organism is not strictly determined by it. Genetic determinism is untenable (see, e.g., Weiss 1973). The goal-direction during the development of an organism is the result of the netted interaction of genome, cytoplasm, and environment.

The question that arises now is the following: how is goal-directedness possible in a complicated network of interactions? The answer was given by cybernetics and automata-theory in terms of feedback-loops (see, e.g., Wiener 1961; Young 1969; Monod 1970; Wuketits 1978). "The nature of feedback is that it gives a mechanism, which is independent of particular properties of components, for constituting a stable unit. And from this mechanism, the appearance of stability gives a rationale to the observed purposive behavior of systems and a possibility of understanding teleology (in the sense of goal-directed behavior)" (Varela 1979, p. 167). Thus, cybernetics and its application to biological phenomena constitutes a breakthrough for biology and many other disciplines. Bateson (1972, p. 470) went as far as to say that besides the treaty of Versailles the development of cybernetics has been the most important event in the twentieth century (from an anthropologist's point of view). General systems theory has gone even beyond cybernetics as far as integration, complexity and generality of systems is concerned. It was developed independently in the USSR (see Takhtajan 1971) and in the West [Bertalanffy (1968, 1975), Laszlo (1972a,b, 1974) and many others; see also the Yearbooks of General Systems Theory]. Naturally, general systems theory is not only of fundamental importance for modern biology (see, e.g., Weiss 1973; Wuketits 1978), but for all aspects of life since the most inclusive system is the whole world and cosmos. Thus Laszlo presents a "systems view of the world" (Laszlo 1972b) and "a systems approach to world order" (Laszlo 1974).

Cybernetics as well as systems thinking have gone far beyond the traditional way of thinking in terms of linear causality since they are based on circular and network causality. It was this new basis that, in terms of feedback-loops and multifactorial interactions, provided an explanation for goal-directed behavior without the direct implication of consciousness. It made possible the construction of machines (servo-mechanisms) that exhibit goal-directedness. These machines, such as thermostats and goal-tracking missiles, behave as if they knew their goal, as if they anticipated consciously their goal. Yet, consciousness is not involved in these phenomena that result from the presence of feedback loops. It has been said, however, that such machines as exhibit goal-directedness are made by us and hence are extensions of our conscious intentionality. This is correct in one sense (see external teleology, Sect. 7.2), but what matters here is that they exhibit goal-directedness because of feedback loops that reside within them. We have also simple machines that do not have such built-in feedback loops. These machines lack goal-directedness as an inherent property, although in the sense of external teleology we have built them in order to reach certain goals.

Machines with built-in feedback loops that lead to goal-directed behavior are of great importance to biology as an analogy for the goal-directedness in living organisms. What Cannon (1936) called the "wisdom of the body," i.e., its goal-directedness in phenomena of homeostasis can be well explained in cybernetic terms without the assumption of intentionality, a vital force, or the interference of supernatural cosmic forces or God. Thus, cybernetics gave a blow to vitalism, spirituality, and theism inasmuch as they were justified by phenomena of goal-directedness which are now explained mechanically (see also Sect. 9.2 on vitalism).

In the preceding discussion of goal-directedness, which does not imply consciousness, I have been referring to individual living systems because as far as such systems are concerned the phenomenon of goal-directedness is accepted by most biologists. Some may use other terms to refer to this phenomenon such as 'aptitude,' 'adjustment,' or 'capacity' (Bernier and Pirlot 1977, p. 70). In any case, the phenomenon is not very controversial as far as individual living systems such as organisms are concerned. However, the situation is quite different with regard to the evolution of individual living systems and groups (taxa) of such systems. Whereas the majority of present-day biologists reject goal-directedness in evolution, some have dared to ascribe goal(s) also to evolution. For example, Teilhard de Chardin (1955) considered humankind as the goal of evolution and assumed that it will evolve toward a further goal, namely, the point Omega at which the separate consciousness of individual men and women will fuse into a collective consciousness so that a result humankind will become much more integrated, harmonious, and peaceful.

Any goal-directedness in evolution is vehemently denied by the Synthetic Theory of Evolution (STE). For example, Mayr (1969), Monod (1970), and Jacob (1970) state emphatically that evolution is the result of random mutations and the selection of organisms and population that exhibit them. If anything has evolved that appears to be the result of goal-directedness, it must be due to pure chance. There cannot be any goal or purpose in evolution because evolution is said to be blind (Monod 1970).

Biologists who argue in this fashion have, however, overlooked that there are also feedback loops in organic evolution. As a result of these feedback-loops a certain directionality may occur during evolution. The kind of directionality I am referring to is well known to paleontologists by the name of orthogenesis or ortho-directionality. Whereas the former term often has been interpreted in terms of intentionality or guidance by a spiritual cosmic force (i.e., teleology in the sense of definitions 2, 3, or 4), the latter term refers to goal-directedness in terms of cybernetics and systems theory (definition 1).

With regard to plants, Meyen (1973, 1978) has shown that the same or similar sets of structures occur in different taxa. He called these sets Repeating Polymorphic Sets (RPS) or simply refrains. It would seem unlikely that such repeating polymorphic sets have evolved independently as a matter of pure chance. Therefore, Meyen interprets them as the result of laws of evolution. Since the feedback loops and the network of the individual organisms may select mutations (see "internal selection," Sect. 8.5.2), they influence the direction of evolution and in this way might extend the goal-directedness of the individual organism into the evolution of taxa of organisms. However, since selection by the environment ("external selection") also plays a role, the internal directionality may be more or less dispersed. This explains why goal-directedness is less obvious in phylogeny than in ontogeny and therefore debated or even denied.

As Waddington (1970, 1975, 1977) and others have pointed out, evolutionary theory has tended to neglect the ontogenetic development of organisms, i.e., the events leading from gene(s) to character(s) and from the genome to the organism. If phylogeny is seen as a succession of (modified) ontogenies, we have a basis for integrating the goal-directedness of the individual living systems as it is exhibited during ontogeny with the more debatable goal-directedness of evolution. If, in addition we shall aim at integrating goal-directedness and goal-intendedness through a cybernetic and systems-theoretic analysis, then we may indeed gain more profound insights. The crux from this point of view is integration.

## 7.5 Function

Function may be related to goal-directedness, but it does not coincide with it. Among its various meanings I want to concentrate on the following two aspects: (1) function as an activity important for survival (definition 4, Sect. 7.2), and (2) function as any activity that is integrated into the whole system (definition 3).

The first of the two definitions "draws attention to the fact that what is under consideration is an adaptation or something which confers an 'adaptive advantage' on its possessor" (Ruse 1973, p. 184). The systems that exhibit this kind of function, which is important for survival, have been termed 'adapted systems' (or by some authors also teleological or teleonomic systems). Examples of such adapted systems are the heart, or the kidney, or the brain. They are considered to be important for survival and hence constitute a function in the sense of a vital function (Woodger 1967, p. 327).

I want to mention briefly two problems that arise from this kind of functional consideration of adapted systems:

a) The functions of adapted systems evolve in time. Is this evolution goal-directed? As pointed out already, most biologists deny goal-directedness in evolution. They insist that even the evolution of the most adapted systems is due to the natural selection of random mutations (see, e.g., Monod 1970; Mayr 1982). However, if we take into consideration that the organism selects mutations due to its organization and thus imposes a certain directionality on evolution we may come close to goal-directedness. As a consequence, the concept of function may be related to that of goal-directedness (definition 1). Since functions are not static, but may also change in time, functions as contemporaneous relations acquire a temporal aspect that, depending on one's interpretation of evolution, may be seen as an expression of goal-directedness. It seems that this possible connection between functions and goal-directedness is one reason why many authors tend to lump functional statements together with teleological (teleonomic) and finalistic statements in the sense of goal-directedness. It is also easy to see that for those who believe that evolution is guided by a supernatural force adapted systems are teleological (according to definitions 3 and 4) or finalistic in a corresponding sense.

b) A second problem that may arise when "function" is defined as a relation important for survival is the following: what is considered "important" for survival? What are the criteria for "important"? It has been said that the heart beats in order to pump blood which among other interpretations can mean that the function of the heart is to pump blood. In this case it is pretty obvious that this function is vital which means that it is important for survival. However, in many other instances it is not so clear whether an activity is important for survival. Hence it is not clear whether it is a function in the above sense. For example, it has been assumed that the heart does not beat in order to produce heart sounds, i.e., heart sounds have not been considered important for survival. Yet, it has been shown that heart sounds perform a vital function (see Bernier and Pirlot 1977, p. 110). In many other cases it is still very difficult to decide whether an activity is important for survival. It may be that an activity is of rather indirect importance. Where then does one draw the line between what is important and what is not? Since living systems are highly integrated systems, all activities may be considered to be important in the widest sense of the word. However, instead of stretching too much the meaning of "important," it may be preferable to resort simply to a meaning of function that does not imply the notion of "important for survival." Then "toutes les activités sont des fonctions" ("all activities are functions") (Bernier and Pirlot 1977, p. 110).

According to definition 3, a function is an activity that relates to the whole network of activities. The network of living systems and its implications for causal thinking have been discussed already in the preceding chapter. In that context the question was how the whole network determines a particular thing or event that is part of that network. In concrete cases of biological research this amounts to the multifactorial determination of a thing or event. In contrast, functional analysis attempts to show how one particular event (i.e., activity) in the network relates to the whole network or a part of the network. An example is how the activity of the hypophysis relates to the activities of the central nervous system, the peripheral nervous system, the muscles of the uterus, etc., of a woman who is giving birth to a child (Bernier and Pirlot 1977, p. 110). It is obvious that these interactions can also be studied in terms of network causality. In fact, as the biologist is investigating the interactions he may be constantly shifting his perspective in order to unravel the complexity of the network. Thus, at one point he may look at one activity as being determined by the network, whereas at another point he may ask how one particular event relates to others in the network. The first point of view shows network causality, the second one the functional aspect. Since the two are related in that they refer to the same network, it is easy to see why some authors such as Bernier and Pirlot (1977) conclude that functional statements can be reduced to causal statements. I think, however, that functional statements entail a perspective of the network that differs from that of the statements in terms of network causality. Hence, I do not see how the functional perspective can be reduced to the other perspective in terms of network causality. Both perspectives complement each other: the functional perspective shows how one activity relates to the whole network or a part of the whole, whereas the perspective of network causality demonstrates how that particular activity is dependent on the whole network. Both perspectives are of heuristic value. If we were to eliminate functional analysis we would deprive ourselves of a powerful method and thus diminish the chances of acquiring a better understanding of living systems. In observing a structure or an event we may ask the question: what is its function in the whole system or in a subsystem? Or we may ask the other question: how is it determined by the network of the whole system? Both questions can be answered only if we understand the organization of the network of the whole system. In that sense both questions refer to the same network and require an elucidation of the same network. It seems that both perspectives could be derived from a knowledge of the organization of the whole network. However, since we do not know that organization completely, such derivation is not in reach or only partially.

In analogy to the notions of network causality and linear causality, one may distinguish between network functionality and linear functionality. In the former, any function is related to the whole network or part of it, whereas according to the latter a function is related to a chain of events. In the simplest case the chain may consist only of one or two elements. For example, the beating of the heart may be related to the circulation of the blood. The causal reduction of this functional statement may seem to be as to state: the beating of the heart is the cause of its circulation. However, Beckner (1969) has shown that the causal statement is not completely equivalent to the functional one because the former says more than the latter.

[Beckner (1969) uses Nagel's (1966) chlorophyll example in his criticism of the supposed translatability of functional statements into causal ones.]

If we understand functionality with regard to the whole network, we obtain a more adequate view of the relations within a system. As more and more inclusive systems are considered, we may end up with the most inclusive system, namely the whole cosmos. In that system, the relations of the stars to the whole system or subsystems may be studied. This means that in that perspective stars also may have functions. Similarly, it may be said that the moon has functions not only with regard to the tides but also with regard to living systems. The functions of the moon are the result of integration between the moon and the biosphere. Those who fail to recognize this integration would tend to deny that functions may be ascribed to inanimate bodies such as the moon. However, if one realizes that a sharp borderline between the living and the nonliving systems cannot be drawn, then functions can be ascribed to any thing or event that interacts with other events in a system or subsystem. I think that to some extent the term network functionality may be substituted by the term 'integration.' Integration can be seen as a relation of one particular thing or event to the whole, emphasizing that it forms an integral part of that whole.

## 7.6 Teleology and Reality

In conclusion, I think that the notions of teleology in the sense of goal-intendedness, goal-directedness, and functionality are useful and adequate to a certain extent inasmuch as they guide biological research and as they present an aspect of nature. In terms of Woodger's (1967) map analogy one could say that there is a certain correspondence between the above teleological and teleonomic notions on the one hand and nature on the other hand. This does not mean that nature as she exists is teleonomic or teleologic in the more highly evolved living systems. According to what has been said, especially in the preceding chapter on Concepts and Classification, concepts are abstractions and as such do not totally correspond with nature. Thus, any scientific and conceptual approach to nature can provide at best a partial correspondence, i.e., a partial truth. Scientific propositions can portray nature only to a limited extent. This applies also to teleologic and teleonomic propositions. They are, however, limited in yet another sense that is related to the general limitation of science but still more specific. Teleologic and teleonomic propositions refer to ends in the sense of contemporaneous ends with regard to functions or in the sense of temporal ends with regards to goals. In the Oxford Dictionary 'goal' is defined as a "boundary" and "limit." The crucial question that we have to ask now if we want to discuss teleology (and teleonomy) as it relates to reality is the following: are there ends, i.e., limits, in nature independently of our teleologic (and teleonomic) interpretation of nature? My answer is no. Let us look at temporal ends, i.e., goals, first. I think it is evident that a goal is never an end in an absolute sense. If state D is the goal, it will always be followed by other states and in that sense it is no end. There are no ends in nature. So what justification do we have to use the notion of goal in science? This has been alluded to already. The goal may be a state that is regularly reached even when there has been diversion in the succession of preceding states. Thus, the goal can be singled out as a state upon which many sequences converge. This is not fiction, but a state of affairs in nature. Furthermore, a goal may be a more stable state than others. This stability again reflects a property of nature that is not purely fictitious. Nonetheless, we have to realize that even the most stable state eventually will be followed by another state, i.e., the most stable state, which at one time may give us the impression to be the end, is no end in an absolute sense; it is at best a temporary end. And all those sequences of events and states that converge upon one state, the so-called goal or end, eventually continue beyond that state toward other states. Figure 7.2 presents in a simple graphic way the goal-directed behavior of three sequences of states from the time $t_1$ to $t_n$. At the time $t_5$ all three sequences converge upon the same state, the goal, which remains stable for a prolonged period of time till $t_{10}$ when the states of the three sequences diverge again from each other to various degrees. As long as we consider these three sequences of events in isolation, the goal-directed behavior is striking and we can see that it refers to something in nature. However, if in a somewhat more realistic view we take into consideration the network of which the three sequences are a part (Fig. 7.2.b), then it is obvious that the goal-directedness of the three se-

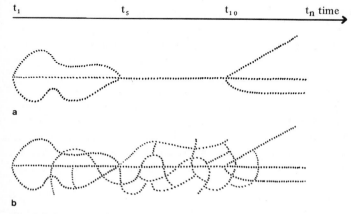

Fig. 7.2. Explanation in text

quences is an abstraction from nature. As pointed out in the chapter on Concepts and Classification, abstraction need not mean that there is no basis in nature. In fact the diagrams of Fig. 7.2 show that there is a basis in nature. But the concept refers only to a part, an aspect, of what is there and therefore is not a complete representation of nature. Nature seems to be neither teleological nor teleonomic. Nonetheless, she is such that teleological and teleonomic concepts may be applied. This still leaves it open whether in the future more adequate concepts than those of teleology and teleonomy may be introduced.

Needless to say, the states or events represented by the dots in Fig. 7.2 are also abstractions from nature. Inasmuch as they are adequate abstractions they also represent an aspect of nature, but not nature as she is (see also Bohm 1980).

Let me now illustrate all this by examples. Riedl (1980, p. 152) mentioned that the goal of the chicken is to lay eggs, and that the goal of an egg is to become a chicken. Although such expressions seem to have a certain justification in biology, they have to be seen in context. First of all, neither the chicken nor the egg is a terminal state in nature but only in our conceptual scheme. Secondly, the egg and all the developmental stages leading to the chicken are part of the network of a more inclusive system. In that inclusive system the succession of states leading to the chicken is an abstraction from the multitude of interactions occurring. It therefore may be misleading when Riedl (1980, p. 152) says that nothing else is built into the egg than the goal to become a hen ("nichts anderes ist dem Ei eingebaut als das Ziel, eine Henne zu werden"). It depends very much on the inclusive system what happens to the egg. A goal as a definite built-in entity does not exist in nature.

As far as a human foetus is concerned we might say its goal is to develop into a mature human organism. Maturity is, of course, no end state in nature because it is followed by other states, eventually by death. If death is considered the end, then the goal of human development would be to die. However, death cannot be a final state either. It is followed by other states about which we know very little. Regardless of whether we believe in some kind of persistence of the soul after death or whether we reject even the notion of soul (which is also an abstraction), we cannot deny that at death a transformation occurs, a transformation in a bodily and/or spiritual sense or in a sense that may better not be expressed in the conceptual dichotomy of the material and immaterial. Many poets and visionaries have been telling us that nothing dies. There is only transformation; hence there are no ends in an absolute sense.

We are so much used to the concepts of "goal" and "end" that we tend to believe eventually that ends and goals actually exist in nature. Another example may illustrate this. As I walk along the beach, I leave footprints in the sand. A little while later as I return these footprints are gone; they have been washed away by a few big waves. Are they really gone? Have they really disappeared so that no trace of them is left? Has their existence really ended? I think nothing could be further from the truth than an affirmation of these questions. Although they have disappeared as the visible depressions in the sand, they have not disappeared completely: they have become transformed. Footprints are depressions in the sand that influence the mutual arrangement of thousands or millions of sand grains. This arrangement of the sand grains constitutes the footprints. As the footprints are "washed away," the arrangement of the sand grains changes to the extent that it no longer satisfies our definition of 'footprints.' If we call this change "disappearance," or "end," or "death," we have not understood much of what has happened. We simply have imposed concepts that give us the impression of a discontinuum where there is none in nature. If the change that occurred is rather drastic and sudden, then these concepts may be adequate inasmuch as they represent an aspect of nature. However, even in cases of adequacy – and especially in those cases! – we should realize that there remains a difference between the description of nature and nature herself. This difference may be of utmost importance with regard to our experience of the world and our quality of living (see, e.g., Watts 1951, 1970; Krishnamurti 1970; Krishnamurti and Bohm 1985).

When we return to the beach and note that our fottprints have been washed away, we might be tempted to say: they are gone and the beach is again just the same as before we left the mark of our footprints on it. As long as we apply the concept of smoothness to the beach, we may indeed state that the beach is again smooth as it was before. However, if we consider the reality of that beach we must admit that it is not the same before our feet touched it and after our footprints have been "washed away." Our footprints have left their mark on the beach even after they are no longer visible because as a result of those footprints the arrangement of the sand grains is different. Inasmuch as our footprints influence this arrangement of the sand grains they leave their mark and the beach will never be the same again. Thus, after you have touched a man or a woman, he or she will never be the same again. After you have looked into someone's eyes, (s)he will never be the same again. After you have said something of importance, your audience or the other person never will be the same again. Concepts that are unchangeable, therefore, cannot express reality.

A clarifying remark may be appropriate with regard to the meaning of "aspect." Figure 7.2b shows that as a more comprehensive system it includes the sequences of Fig. 7.2a. What has been represented in Fig. 7.2a therefore can be characterized as an aspect of the whole of Fig. 7.2b. It is important,

172

however, to realize that the difference between Fig. 7.2a and Fig. 7.2b is not only a matter of addition. The whole context determines that which is singled out in Fig. 7.2a. Hence, what is shown in Fig. 7.2a does not really exist as such as it is represented in isolation. This means that if Fig. 7.2a is considered to be an aspect of nature it is so in a very indirect sense.

What now is the status of purpose in terms of function? Inasmuch as a function is a concept, i.e., an abstraction, it also removes us one step from reality. If functionality is understood in terms of the most inclusive whole it may refer rather adequately to reality. However, even then it presents, as we have seen, just one perspective of the whole. Any perspective of reality, as the term perspective implies, must be dependent on the observer or knower and therefore is limited. Hence, my conclusion concerning the reality of functions is the same as with regard to goals: functions are neither absolutely fictitious, nor absolutely real. Our goal is to invent notions of functionality that are increasingly adequate and to apply these notions in such a way that a maximal correspondence of these notions and the natural system is obtained. I think that the notion of integration is very promising and may perhaps constitute a more general and possibly more adequate concept than that of function. It is of great usefulness in modern biology. However, even with regard to the notion of integration, which refers to the integration of entities such as things and events or processes, we have to realize that in reality entities are not there; they themselves are abstractions, i.e., they do not exist as such. How could something that is not absolutely real have an absolutely real function!

Experience in terms of functionality and integration can also be limiting, especially when it is restricted to a small subsystem whose integration into more inclusive systems is not recognized. With regard to human relations such restriction may lead to much suffering and frustration because of the unrealistic belief that the subsystem is autonomous. It seems that many romantic lovers tend to see their relationship as an autonomous whole and disregard quite unrealistically that it is embedded into a much more inclusive whole. The discrepancy between the imagined autonomy and the real interdependence may be the source of much suffering and may destroy eventually the most beautiful relationship. The unrealistic attitude may also show up when the lover is "lost." Then the "deserted" partner may find himself or herself in a cold and barren world, although (s)he may be surrounded by the most beautiful flowers and trees, a sparkling green sea and a glorious blue sky. The boundless sorrow results from seeing the function of one's whole existence solely in relation to the beloved person and thus isolating in one's mind one relation from the context of the more inclusive whole. Novalis experienced such despair and expressed it in the most romantic poetry. He finally transcended his blindness and loneliness by seeing through the imagined limits of his love relation. During this transformation of his vision his love became elevated to more and more inclusive religious experiences. In a similar way, in Sufism erotic love may acquire truly universal dimensions. Thus, the notion as well as the experience of functionality within self-imposed unrealistic boundaries is transcended: love becomes whole and holy.

If purpose is defined in a wide sense so that it comprises goal-directedness as well as functionality, I conclude that in that sense there is no purpose in nature. In other words: purpose is not absolutely real. That does not imply that purpose is purely fictitious or a totally arbitrary concept projected into nature. I think that there is an aspect of nature that is rather adequately dealt with in terms of the notion of purpose, i.e., teleonomic and functional propositions. Since that aspect is part of nature, it might be said to be real. In this weak sense of the term real, purpose may be real. Only in this sense can I accept the statement that "there are indeed goal-directed systems in nature" (Ruse 1973, p. 190). However, if real is that which exists independently of our knowledge of it, then purpose including goal-directedness is not real because it is partially a function of our conceptualization. "'Purpose' appears as the universe is dissected" (Bateson 1979, p. 229).

The conclusion that nature, undissected by our thought, has no purpose is not at all negative. It simply means that nature is infinitely more than any conceptual description could indicate. She is neither negative nor positive, neither bad nor good. And in that sense one might even say that she has neither purpose, nor no-purpose. This, of course, is only a different, perhaps more careful, way of expressing what I tried to convey in the preceding section.

Having discussed purpose in the sense of goal-directedness and function, I shall now turn to goal-intendedness, i.e., goals or purpose as we know them in our life and the lives of other human beings. Is not it real that our lives are shaped by the goals we are pursuing? This is a very difficult question to answer. In one sense it appears to be a truism because we can see that our lives may change drastically if we change our goals. If a businessman suddenly decides to give up his business in order to pursue the goal of a university education in biology, he will experience very dramatically how this new goal affects his life. Yet we may question whether human goals are real as goals. As Mohr (1977) pointed out, goal-setting requires consciousness and free will. Whether free will exists is debatable and it seems that there will be no last word on this issue either. As I pointed out above, our behavior is influenced to a great extent by our subconscious. Hence, the role of consciousness in goal-setting is also debatable. This means the existence of goal-intendedness is questionable at least to some extent even in the human species. Existence here is again meant in an absolute sense. I think one cannot deny that the concept of goal-intendedness applies to human thought and action and thus represents at least an aspect of human behavior. However, whether it really exists is a different matter. Mohr (1977, p. 94) also considers it "doubtful whether 'in reality' we are able to act teleologically."

These statements do not appear to be very helpful as far as the conduct of our lives is concerned. In our daily life we seem to be confronted with the experience of choice: should we pursue this goal or another goal or no goal at all? Many people suffer very much under the burden of this experienced choice. We may go further, though, and ask what can or should be our attitude toward goals in general. Should we believe Riedl (1980) and others who claim that it is perfectly natural to lead a life that is fundamentally determined by goals and thus obtains a purpose? Or should we look upon this view as the result of social conditioning by a culture that is possessed by teleological ideas and ideologies? Many sages in the East and West have told us that "man's identification with the choice-generated world of ideation, illusion, and make-belief" (Deshpande 1978, p. 4) is at the root of our problems. What can be done in this situation? Deshpande (1978, p. 5) describes Patanjali's Yoga sutras as illuminating "a state of being in which the ideational choice-making movement comes to a standstill." It seems that many practioners of enlightened ways of living have reached this state that is far beyond the goal-ridden anxieties of the more or less neurotic slaves of many modern societies.

More and more people, especially among the younger generation, seem to realize that slavery to goals and ends is not the way to a happy and fulfilled life. They are aware that goals and ends impose unnecessary limitations. The more we identify with our goals, the narrower the focus of our life becomes. In the extreme the relation to the whole is more and more lost and the result of the unnatural one-sidedness and narrowness is perversion, destruction, and much suffering. For example, the person who does not reach his goal to buy a car or to obtain a Ph.D. may feel depressed and miserable and thus may miss all the beauties of life around him. The narrow-mindedness of goal-dominated behavior may lead to blindness that in turn may have catastrophic consequences.

Many young people have dropped out of our goal-oriented society in order to lead a simple life in harmony with nature. Are they not following a goal too? Some of them may without realizing it, others may transcend the initial goal and thus reach "a state of being in which the ideational choice-making movement comes to a standstill" (Deshpande).

Whether one drops out of society or stays inside is a personal decision. It seems to me what matters is not so much where one lives, but how one lives, although some environments are more conductive to an enligthened life than others. Important is awareness of the relativity of the goals. A person who has this awareness may pursue goals and in fact may live with others in a goal-oriented society. Yet that person is not the slave of goals because to him or her goals have become transparent as something relative in the whole. The life of that person is whole and not narrowly conceived and distorted by a one-sided goal.

I think the reason why many people still survive relatively well in our goal-ridden society is that they are still able to "drop out" periodically and thus restore the natural equilibrium. Thus many activities of our leisure time compensate for the loss of wholeness in the narrow-minded pursuit of goals. For example, dancing and playing may have a spontaneity that surpasses any rigidity of goal-directedness and purposeful action. In this sense we may understand Jesus' words: unless you become like children you shall not enter heaven. To Yogis and other sages, the highest state of being is totally spontaneous and therefore without purpose. It is the highest state because it is in complete harmony with nature which is also without purpose.

Having no purpose is often seen as negative and is experienced as a restriction. However, the opposite may be said: having a purpose means limitation by an end that is singled out from the continuity of being. Furthermore, the selection of one end restricts our perception of the world because it is limited to one perspective. Accordingly, our actions become channeled into one narrow way. The wholeness is lost in experience as well as in action.

So if goals are irrelevant as ultimate guidelines for our life, where should we be going? This question is also irrelevant when goals have lost their grip on us because we are going nowhere (see, e.g., Trungpa 1981). Since nature of which we are part is not composed of absolutely distinct entities, i.e., things, there are no things that could constitute ends for our behavior. If we want to go somewhere, we first have to abstract an end and then direct ourselves toward that end. But in this process we limit ourselves and lose the whole. Only in going nowhere, we are everywhere. Alan Watts made a film in which he contrasted the behavior of two brothers. One of them went around the world rushing from one goal to another, gaining many new experiences, but without really seeing. The other brother stayed in his garden and saw.

Going nowhere and seeing is difficult for most of us. Goals of all sorts have too much a grip on us. However, many of us at least occasionally have experiences and insights that are beyond the ordinary purposiveness. If we do not run away from such happenings but instead open up more and more, goals may gradually lose that power over us and we may come to "a state of being in which the ideational choice-making movement (toward goals) comes to a standstill" (Deshpande 1978). Then, one can be and see.

In order to reach that state much discipline may be required. The resulting striving may again reinforce purposive behavior. Note the preceding use of the teleological phrase 'in order to'! However, this goal-intendedness may be only a means to reach a state that is beyond it. Thus the initial purposiveness may be compared to a boat that is used to reach the other shore of a river. When one is there, one does not need the boat anymore. It is left behind and thus becomes irrelevant.

I think that there is a close connection between goal-intended behavior and stress. As we rush from one goal to another we suffer more and more from stress and it becomes increasingly difficult to relax physically as well as mentally. Goals may be projected even into relaxation. However, as long as we pursue relaxation as a goal we cannot truly relax. Complete relaxation is letting go of everything, even the goal to relax.

At this point I want to refer briefly to the teleology of certain philosophers and theologians who claim that nature is fundamentally purposeful, that there is a universal purpose (of a Creator) underlying everything, although it may be totally or partially hidden to us (see, e.g., Teilhard de Chardin 1955). If this claim is grounded in a profound experience one can, of course, challenge it only if one shares this experience. If it is predominantly an intellectual speculation of a mind that has been conditioned by a goal- and purpose-oriented culture the matter is different. In that case I think the postulated purpose in nature is mainly our projection. However, as pointed out, this projection is not totally arbitrary, but has a certain empirical basis. I think it is because of this empirical basis and our social conditioning that grand schemes of purposiveness in nature still have a great appeal to many thinkers.

Riedl (1980) also believes that there is purpose in nature. However, to him purpose seems to be mainly constituted by functionality. He believes that due to natural selection the evolution of humankind has led to a stage where nature has become aware of herself. This belief has been expressed in different cultures and in different ways. Now Riedl goes one step further in making the additional assumption that it is Western culture in its scientific culmination that is the pinnacle of evolution and hence comes closest to a portrayal of reality. In his Fig. 56 (1980, p. 168) the European is on the top of the evolutionary progression that took its origin from primitive animals. Riedl does not justify this implicit superiority of the European and Western mind. Perhaps he considers the high degree of scientific development and intellectual capacity as a sign of superiority. But who can substantiate that the Western intellect provides a better understanding of reality than the intuition of sages in the East and West?

It seems to me that Riedl's outlook and consequently his conclusions are biased in favor of Western thinking and thinking (i.e., cerebralization) in general. Riedl does not sufficiently take into consideration that there are other cultures and a counterculture in the West (see, e.g., Roszak 1969, 1973) that challenge fundamentally his view of things and the underlying state of consciousness. Whereas to a Western thinker the head and the brain are the most important human organs with regard to the understanding of nature, in other cultures the focus may be on the whole mind-body or as in Zen Buddhism at the hara which is a region between the navel and the genitals. Furthermore, whereas to a Western thinker the intellect, the mental, and the spiritual in a restricted sense ("das Geistige" in the German language) are in the foreground, to many sages breathing and the movements of the body-mind are central. In Zen Buddhism mind is transcended to no-mind. In that state reality is seen in its suchness.

I think that one can work as a scientist in a purely Western context. However, if as a "hypothetical realist" (Riedl 1980) one has the ambitious goal to penetrate the core of reality, a biased pro-Western outlook will not be a sufficient basis for such an undertaking. If we try to look at the question of teleology in as comprehensive a way as possible by considering not only results of science and philosophy but also the variety of experiences in different states of consciousness, we must at least question whether there is purpose in nature. We may act according to goals, but we may also transcend goals and purpose. One might argue, then, that our purpose is to become aware that there is no purpose in nature including ourselves. However, since the awareness of no-purpose is beyond purpose and any conceptual comprehension of the world such a statement is meaningless, and the purely intellectual discussion is just a quibbling over words and concepts.

Even for those who realize this it is not always easy to convey the no-purpose of nature and life in verbal communication. Thus, Lin Yutang (1938, p. 119) stated that the purpose of life "is the true enjoyment of it." He added, however, that this is "not so much a conscious purpose as a natural attitude toward human life." He warned that "philosophers who start out to solve the problem [i.e., pseudo-problem] of the purpose of life beg the question [of living] by assuming that life must have a purpose." And he added: "I think we assume too much design and purpose altogether" (Lin Yutang 1938, p. 119). I have quoted specifically Lin Yutang as a representative of the Chinese culture which in many respects differs radically from Western cultures (see, e.g., Nakamura 1964). Lin Yutang emphasized that in China living as enjoyment is considered more important than abstraction and intellectualization about life as in science, philosophy and to some extent even in poetry. I realize, of course, that China too is more and more influenced by Western mentality.

I suggests that the reader look upon my conclusions in the spirit of Dada. As Feyerabend (1975, p. 21) put it, "a dadaist is convinced that a worthwhile life will arise only when we start taking things lightly and when we remove from our speech the profound but already putrid meanings it has accumulated over the centuries ('search for truth'; 'defense of justice'; 'pas-

178

sionate concern', etc., etc.)". Since I have not been able to avoid all of those profound and possibly putrid concepts, I hope that wherever they have slipped in the reader will take them with a smile.


## 7.7 Summary

*Introduction.* Teleology refers to goals, purposes, and functions, which in turn describe or relate to the most complex biological phenomena such as homeostasis, self-regulation, adaptation, selection, integration, organization, programs, and feedback. Because of the complexity of the subject matter, the semantics of teleology is in a confused state. A number of terms are used to refer to an even greater number of concepts which are mutually exclusive or more often overlap more or less with each other.

*Terminology.* The word 'teleology' is derived from the two Greek words: telos (end, goal) and logos (reason, doctrine). Thus, its literal meaning is the logos of ends. It has been equated to goal-intendedness, goal-directedness, purposiveness, functionality, adaptedness, finality, etc. Internal and external teleology have been distinguished. The term teleonomy has been introduced to exclude conscious intentionality and the direction of processes by a god, a creator, or spiritual forces. The concept of function has also been defined in rather different ways. Six definitions are distinguished.

*Goal-intendedness.* Teleology as goal-intendedness is discussed on the basis of Mohr's model according to which the following elements are required for goal-setting: consciousness, rationality, free will, values, propensity structure, and experience. All of these elements are discussed critically. It is doubtful – also according to Mohr – whether this kind of teleology exists even in the human species. Whether it occurs in at least the more highly evolved animals is even more conjectural. Relaxations of Mohr's model are discussed. The most relaxed form would require only experience, propensity structure and degrees of consciousness. The resulting concept of teleology might be applicable to animals and plants. However, it is highly questionable whether the existence of consciousness in animals and plants can be investigated scientifically and it is also debatable whether consciousness can be treated as a fuzzy set.

*Goal-directedness.* Cybernetics, which according to Gregory Bateson constitutes the most important development in the twentieth century besides the treaty of Versailles, has provided models of feedback that made possible the construction of servo-mechanisms whose behavior is goal-directed such as, for example, that of thermostats or goal-tracking missiles. Organic goal-directed phenomena, such as homeostasis and self-regulation, can also be explained in a cybernetic way without recourse to consciousness or spiritual forces. Although it is usually acknowledged that goal-directedness in a cybernetic sense occurs during ontogeny, it is still very much debated whether this notion applies to phylogeny as well. The synthetic theory of evolution (STE) rejects any form of goal-directedness. However, modified and extended versions of STE that take into consideration internal selection provide a framework for the conception of cybernetic goal-directedness in phylogeny. It is a great challenge for modern biology to integrate goal-directedness and goal-intendedness with regard to ontogeny and phylogeny.

*Function.* Two meanings of the concept of function are discussed: (1) Function as an activity that is necessary for survival, and (2) Function as an activity that relates to the whole system. The second definition appears to be more appropriate and of more general usefulness. As is the case with regard to causality, functionality can also be conceived of in a linear or a network fashion. Evidently, network functionality is a more adequate concept for complex living systems than linear functionality. Both network causality and network functionality refer to the same network and system. However, the perspective is different: network causality describes how the whole network determines a single structure or process, whereas network functionality refers to the role that one structure or process plays within the whole network. Hence, network causality and functionality represent complementary views of the same system. They deal with the interactions in an integrated system, i.e., the organization of that system. To some extent the term network functionality can be replaced by the general term integration. Since living systems and the so-called abiotic environment are integrated with each other, functions may also be ascribed to physical objects such as the moon.

*Teleology and reality.* There are no ends and hence no goals in nature. There is only transformation. Functions are also abstractions and in this sense, like goals, do not exist in reality. Therefore, if purpose is defined in a wide sense comprising goal-directedness and function, it is concluded that there is no purpose in nature. As Bateson (1979) pointed out, "'purpose' appears as the universe is dissected." This conclusion is not meant to be negative. It simply means that nature is infinitely more than any conceptual description could indicate. She is neither negative nor positive, neither bad nor good. In this sense it may be appropriate to say that she has neither purpose nor no-purpose. Nonetheless, the notions of purpose, function, and goal-directedness seem to portray an aspect of nature and in this sense are useful concepts.

It is debatable to what extent, if at all, we behave teleologically in the sense of goal-intendedness. In any case, attachment to goals does not seem to be the way to a happy and fulfilled life. It leads to an unnatural narrowness and one-sidedness that may entail much suffering. Fortunately, more and more people are becoming aware of this. A remarkable transformation of life-style is happening as a result of increasing awareness.

# 8 Evolution and Change

"Everything flows" (Heraclitus)
"Any describable event, object, entity, etc., is an abstraction from an unknown and undefinable totality of flowing movement. ... Knowledge, too, is a process, an abstraction from the one total flux, which latter is therefore the ground both of reality and of knowledge of this reality" (Bohm 1980, p. 49)

## 8.1 Introduction

Bunge (1979a, p. 331), a theoretical physicist and philosopher of science, distinguishes five kinds of questions (the five W's) that scientific research tries to answer: What? Where? When? Whence? Why? Simpson (1964, p. 111), an evolutionary biologist, suggests that biological questions can be reduced to the following simple and colloquial forms: What? How? What for? How come? "What" questions are asked in descriptive biology at all levels of organization. An example of an answer to a "what" question is: this is a chromosome. "What" questions and their answers were the subject of Chaps. 3 and 4. "How" questions are characteristic of experimental biology, again at all levels of complexity. They imply linear, circular, or network causality. For example, an answer to the question "how does an organism develop?" can go beyond the mere description of development inasmuch as it presents an analysis of causation and/or determination. I have dealt with "how" questions and their answers in Chap. 6. The "what for" question refers to function or purpose (see Chap. 7). "What for is the eye?" is answered by stating that it is for vision, i.e., its function (or purpose) is vision. Both the "how" and "what for" questions can be subsumed under a "why" question. However, in that case the "why" has two different meanings. In the first place it refers to the causal and/or deterministic aspect, in the second to the teleological aspect. As I pointed out in the preceding chapter, these two aspects refer to the same network or system from complementary points of view. When students ask "why" it is often not clear which of the two aspects they have in mind. Finally, the "how come" question refers to the historical aspect, i.e., evolution or origin. "'How come?' means how did this originate? What course has it followed through time and what were the causes of that historical development" (Simpson 1964, p. 112). For example, according to the selectionist view, the answer to the question "how come the giraffe has a long neck?" is that variant ancestral forms with longer necks have been selected. A more complete answer to the "how come" question entails answers to the other three kinds of questions. Thus, the "what" question is answered with reference to "giraffe" and "neck." The two kinds of "why" questions are answered in terms of the function of the long neck and the causal and/or deterministic analysis of its ontogenetic development. With regard to causation, Mayr (1961) distinguished proximal and evolutionary causation. Answers in terms proximal causation (which entail "how" questions) are provided through the causal analysis of ontogeny, whereas answers in terms of evolutionary causation (which are based on "what for" and "how come" questions) are given by evolutionary theory. The latter comprises the former, since phylogeny is a succession of (modified) ontogenies.

Most biologists agree that the theory of evolution is the "single most encompassing biological theory" (Dobzhansky et al. 1977, p. 505). This comprehensiveness of the theory is reflected in the questions asked. For a further elaboration and application of the four kinds of questions to other biological disciplines see Simpson (1964, chapter 6). The criticism of causality and teleology in the preceding chapters should, however, be kept in mind. It may allow us eventually to step beyond the confines of the above kinds of questions and answers (see below the section on evolutionary theory).

## 8.2 Universality of Change

Evolution refers to change. Hence I shall begin with a discussion of change. Change is universal, i.e., everything changes. Galaxies, stars, planets, mountains, oceans, ecosystems, plant, and animal species, men and women, institutions, nations, cultures: all change.

Heraclitus, the Greek philosopher, knew long ago that everything flows ("panta rhei"). As pointed out already, we never step into the same river twice because the river has changed by the second time, and we have changed too. Modern biology has demonstrated that most of the cells of our body are continually replaced by new ones and the composition of the same cell changes rapidly. Hence, in a strict sense, it is never the same cell, nor do we remain the same person.

Change may be rapid or slow. When it is very slow we gain the false impression of permanence. For example, today I may appear to be the same as yesterday, because changes that have occurred since yesterday are not very striking. However, everybody would admit that I have changed during the last 20 years. For example, 20 years ago I did not yet have grey hair. Yet, in some respects I may still be the same as 20 years ago. For example, I may still be romantic as I used to be. In that sense there has been no change. However, I think that this lack of change is only apparent, not real. It is due to abstraction. We single out one general feature. This feature in its general (abstract) aspect remains the same, yet more concretely there are differences. The romanticism is not the same anymore. It has changed too, although the behavior may still be classified under the general category of romanticism. The question is whether this classification indicates permanence at least at some point(s). To answer this question, let me refer to a point within a system of a space-time extension. If we assumed that such a point would retain exactly the same relative position over a certain period of time, could we then conclude that the position of this point remained unchanged? We could say yes in terms of the same relative position, and we may have to say no with regard to the absolute position. Furthermore, we may have to say no because of change in the context. Since the context of the point has changed, it is not really the same point (unless one subscribes to an atomistic world view). As the whole system of which the point is a part changes, the elements are no longer exactly the same. Hence, from this holistic viewpoint permanence does not exist.

Although change appears to be universal, differences between slow and rapid changes may be very striking. Consequently, one can contrast so-called stable states or equilibria with labile ones. The former undergo no noticeable change at the level of the whole system, whereas the latter change rapidly. The apparent antithesis between structure and function (= process) is the result of an extreme difference in the speed of change. The component parts of structures may undergo rapid change (e.g., metabolism). We refer to this change as the function of a structure. The structure as a whole appears to remain relatively stable; however, it changes very slowly and thus constitutes also a process. From this perspective, everything is process, i.e., change (see, e.g., Whitehead 1920, 1929; Woodger 1967, p. 330).

The investigation of differing rates of change can lead to interesting theories. René Thom (1975, 1983; see also Zeeman 1976; Saunders 1980) has approached this question from a global point of view and has developed the mathematics to deal with the whole continuum from very rapid to very slow changes. He has, however, focused his attention especially to the rapid changes which he termed 'catastrophes.' Hence, the name of his theory is 'catastrophe theory.' It has been applied with some success to many fields of enquiry ranging from geology and biology to linguistics and sociology. From an epistemological point of view it is interesting to note that Thom's theory is based on seven elementary catastrophes. Through these elementary catastrophes the dynamic description of the universe is possible. Yet the kinds of catastrophes are unchanging like eternal forms. The question that emerges is: do we require some firm unchanging framework to describe change? If so, is this framework only a methodological tool, or does it indicate that there might be something in reality that does not change?

A look at the history of biology with regard to change and permanence is revealing. Change is often equated with evolution or dynamics, whereas permanence stands for fixism or a static view. Pre-evolutionary biology, such as the systematics of Linnaeus, is often characterized as static in contrast to evolutionary biology which is said to be dynamic. Such a general characterization is, however, not sufficient since pre-evolutionary biology is neither completely static nor is evolutionary biology totally dynamic. Pre-evolutionary biology is static with respect to species. According to Linnaeus there are as many species as were created by God. One species cannot become transformed into another. Hence, species as species do not change. However, within one species variation, i.e., change, may occur. Linnaeus as well as other pre-evolutionary biologists were acutely aware of intra-specific dynamics and accepted its fluidity. Yet they set limits to change at the boundaries of species. Evolutionary biology recognized the unnaturalness of those limits of species. Thus, evolutionary biology is clearly more dynamic than the pre-evolutionary fixism. However, in evolutionary biology there are also limits to change. According to the principle of uniformitarianism the general conditions and principles of evolution do not change during evolution. The basic laws underlying evolutionary change are considered immune to change. Laws in general are thought of as invariant, i.e., unchanging. Thus, stating that the aim of biology is to discover laws governing living systems, is the same as saying that biology aims at discovering invariance, i.e., that which does not change. In other words: the aim is to describe the static and the fixed in life. One could almost say that in this perspective the dynamic aspect of life is only a means to discover the fixism of life.

The question that arises again at this point concerns laws and order in nature and particularly in evolution. Are there really invariants in nature or is this thinking in terms of invariants mainly due to an urge for stability, an urge that may be the result of cultural conditioning and reinforcement? If Heraclitus and other visionaries of a totally dynamic universe are right in claiming that everything flows, then the so-called laws also change and thus cease to be laws in the strict sense. They might change at an extremely slow rate or in such a way that they may still "work" as approximations to a fluid universe, which may not be totally erratic but also not essentially invariant.

Woodger (1967, p. 449) wrote: "If we conceive the world as 'governed by unchanging laws' it is difficult to see how evolution is possible. For, ... evolution appears to require *change* in the 'regularities of change', or at least the coming into being of *new* types of change. And these, on their first appearance, will be *unique*. But if evolution has consisted in a succession of unique changes then it is clearly meaningless to speak of a 'law of evolution' or of a 'cause of evolution' because ... causal laws and the causal postulate do not deal with unique changes" (concerning uniqueness see below, Sect. 8.3). Beatty (1980) also emphasized that laws may be evolving, which means that they are no longer laws in the strict sense. While pointing out exceptions to Mendel's first law and the Hardy-Weinberg law, he adds that the regularities on which these laws are based may change in the future. "Even if there were presently no exceptions to Mendel's law, its truth would be contingent upon the evolutionary status quo. But the evolutionary status quo is hardly permanent" (Beatty 1980, p. 410).

In the *Book of Changes (I Ching)* and the various commentaries on it (e.g., Wilhelm 1960) we find references to laws of change, but we also find them transcended as in the following paragraph (R. Wilhelm's edition, p. 348):

"Alternation, movement without rest, flowing through the six empty places, rising and sinking without fixed laws, firm and yielding transform each other. They cannot be confined within a rule. It is only change that is at work here."

A number of sensitive philosophers and visionaries have been telling us for a long time that life is not rigid and that therefore a biology that insists on static principles in life cannot really understand it. Bergson (1907), for example, stressed the fluidity and creativity of evolution. He therefore thought that evolution cannot be captured by science and thought, which, according to him, can only comprehend the rigid, static, and fixed.

It is understandable that Bergson's ideas are unpopular among most evolutionary biologists. Anyone who clings to law and order and takes comfort and security in a permanence must feel threatened by someone who denies the reality of such permanence and security. Furthermore, the denial or even doubt of absolute lawfulness in nature is often felt as a frontal attack on science. I have already pointed out that this need not be so interpreted, although it does not make life easy for the order-oriented scientist (see, e.g., Lewontin 1966; Stent 1978; Wimsatt 1980; McIntosh 1980/82, 1984; May 1981).

What I think is needed is an increasing adaptation of our scientific models and theories to nature. In this respect there is ample opportunity for evolutionary thinking to improve. I will give only two examples that may illustrate how modern evolutionary thinking can become more dynamic, i.e., closer to nature, than it is at the present time.

The first example concerns evolutionary plant morphology. In this discipline, static concepts of the pre-evolutionary era are still used, although in a somewhat different way (see Chap. 5). Probably the same criticism could be leveled at animal morphology. The approach is often too categorical, too rigid, with implications of essentialism, which entails the existence of eternally unchanging forms. If these implications are removed either totally or at least partially, we render evolutionary morphology more dynamic (Sattler 1984).

The second example concerns the framework of evolutionary theory. Schram (unpublished) has pointed out that this framework is still based on the notions of absolute time and space, notions that have become transcended in physics long ago through the theory of relativity, which postulated a space-time continuum. The adoption of this relativistic framework for evolutionary theory would make it less static and might constitute a considerable improvement. The nonequilibrium theory of evolution (Wiley and Brooks 1982; Brooks and Wiley 1984) also emphasizes dynamics.

Many other changes of evolutionary thinking might be possible. However, they may be beyond our present imagination and thus we will have to await the vision of the genius and hope that the scientific community will not reject the more dynamic because of its urge for security.

In many books the advent of Darwinism is celebrated as one of the greatest discoveries of humankind. One cannot deny that it was an important event in the history of biology. However, one should not forget that in other cultures a dynamic attitude (which is evolutionary in the broadest sense) has been natural for millenia. One of the oldest books written in China is the *I Ching*, the *Book of Changes*. As the name indicate, change, in a very fundamental sense, is the basic theme of this book. Zen, which developed in China and Japan, is totally fluid, dynamic and spontaneous. Biology as a science will not be able to achieve such total dynamism because it may have to be based on thought, which is discursive and static to at least some extent. But who can foretell to what degree of flexibility, fluidity and dynamism future developments may lead us and what will we be able to glimpse at the most daring frontiers of science?

## 8.3 Historicity

Is historical science possible? And if so, how? These are controversial questions (see, e.g., Goudge 1961; Simpson 1964; Ruse 1971; Hull 1974; Mayr 1982). At one end of the spectrum of opinions are those who think that historical science does not differ from nonhistorical science in any important respect (e.g., Ruse 1971). At the other end of the spectrum, historical science is viewed as so strikingly different from nonhistorical science that it may be questioned whether it should still be classified as science.

Major issues at stake in this controversy over historical science are prediction, explanation, and uniqueness of events. With regard to the question of prediction in historical science, most authors agree that it is difficult to make predictions. But whereas some tend to think that this is only a practical difficulty due to insufficient knowledge of important parameters, others maintain that there are reasons that make prediction difficult or impossible in principle (see, e.g., Simpson 1964, p. 138, and below in Sect. 8.5.3).

Concerning explanation, one question is whether the covering-law model applies. Those who answer this question affirmatively accept the existence of relevant laws because such laws are a prerequisite for an explanation in terms of covering laws (Ruse 1971). Others like Goudge (1961) who deny the possibility of explanation by subsumption under laws have proposed an alternative mode of explanation for historical science, namely, narrative explanation (see Chap. 2). Such explanation requires only a knowledge of antecedent events, not laws. Thus, the fact that I am writing this book is explained by a number (or even the totality) of events preceding this activity.

A third major issue at stake underlying the two preceding ones is that of uniqueness. Are historical events unique and, if so, does this preclude any lawfulness and scientific explanation in terms of the covering-law model? I think that events are unique, especially if they are considered in their context. As Stent (1978, p. 219) put it, "every real event incorporates some element of uniqueness." Thus, each sunset is different because the clouds, the air, and many other aspects are different. The important question to ask now is the following: is it possible, in spite of the uniqueness of events, to single out similarities and regularities that would make possible an explanation or even prediction in terms of laws or rules? An answer to this question is not easy to give. It is interesting that those who have a "cool," abstract mind (i.e., the "tough-minded") tend to favor lawfulness also in historical science, whereas the "tender-minded" whose prime concern is the detail of particular situations are overwhelmed by the uniqueness (Maslow 1966, p. 93).

Two levels of uniqueness can be distinguished: the conceptual and the experiential (nonconceptual) levels. At the conceptual level, the unique event is described in terms of concepts. An example similar to the one given by Hull (1974, p. 98) is 'the election of the first Polish pope' (in Hull's terminology, this is a necessarily unique event because repetition is in principle impossible). Although unique as an event, it is described in terms of general concepts such as "election," "pope," and "the first." We cannot categorically exclude the possibility that one day we might come up with lawful generalizations concerning elections, popes, or first kinds of events. In that case the unique election of a Polish pope could be explained in terms of the covering law model as in nonhistorical science. This is not an explanation of the unique event as such, but of some of its aspects that can be abstracted, conceptualized and placed in lawful connections. Inasmuch as these aspects are part of the unique event, we might say that explanation of unique

188

events is possible; but when we refer to unique events as such in their uniqueness, explanation in terms of the covering-law model is not possible. Much of the confusion concerning the explanation of unique events results from a failure to distinguish the uniqueness as such from aspects of uniqueness that can be described in terms of concepts that form part of law statements.

At the subjective, experiential level unique events may not be conceptualized. In that case it is impossible to relate them to scientific laws and to provide explanations in terms of the covering-law model. If we want to analyze such events scientifically we have to abstract from them and in doing this we lose much or all of the uniqueness of the event. Such loss is particularly obvious when a deeply felt experience is made the subject of science. For example, listening together to a piece of music can be an extraordinarily unique experience for two lovers. Any analysis of this event would be intuitively unsatisfactory or incomplete to them because the uniqueness of the event would be lost in this way. Uniqueness here implies richness, fullness, wholeness, and mystery. Since the words wholeness and holiness have the same root (see Bohm in Wilber 1982, p. 53), we may also say that the holiness of the unique event is lost [see also Riedl (1980, p. 81) who refers to the loss of individuality as the destruction of the most humane].

Uniqueness as such and scientific analysis exclude each other because the latter concerns aspects of events and requires their repeatability; uniqueness as such cannot be repeated. One can experience the uniqueness of events or one can focus one's attention on abstracted generalization. In different cultures we may find a varying emphasis on one or the other. Jung (1967) contrasted Western and Chinese mentality in this respect. He wrote (1967, p. XXIII): "While the Western mind sifts, weighs, selects, classifies, isolates, the Chinese picture of the moment encompasses everything down to the minutest non-sensical detail because all of the ingredients make up the observed moment." Needless to say, no culture would restrict itself completely to one or the other activity. We find experience of uniqueness in Western culture as we find abstraction and lawfulness in Chinese culture. Thus, Wilhelm (1960, p. 23) wrote in his commentary to the *I Ching*, the *Book of Changes*: "Change is not something absolute, chaotic, and kaleidoscopic; its manifestation is a relative one, something connected with fixed points and given order."

## 8.4 Evolution and Progress

The term 'change' designates alteration. 'Evolution' can be used as a synonym to 'change' and 'alteration.' In this sense evolution is simply transformation (Zimmermann 1968, p. 20; Wuketits 1978, p. 153), and the term may be applied to living as well as nonliving systems. For example one may refer to the evolution of $CO_2$ during a reaction in a living or a nonliving system.

Often, the term 'evolution' is used with a narrower connotation: either as change that occurs in a certain direction, or change that leads to progress. "The concept of 'direction' implies that a series of changes have occurred that can be arranged in a linear sequence, with respect to some property or feature" (Dobzhansky et al. 1977, p. 508). Directionality may imply irreversibility of evolution. Even if the trend of change were reversed, it cannot lead to the initial state because the context would then be different. Thus, a reversal of the directionality of change can at best approximate the initial state. Only very simple processes such as the mutating back of a gene to its former allelic state can be considered as a reversible change provided it is taken out of context.

Directional and irreversible changes occur in both living and nonliving systems. The second law of thermodynamics refers to irreversible and directional changes, namely the increase of entropy in a closed system. In organic evolution, the phenomenon of directionality is exhibited in evolutionary trends such as, for example, the well-known skeletal sequence from early fossils of horses to the living horse (see, however, Hitching 1982, p. 28).

"Progress" implies more than directionality. It requires that the directional change leads to "a *betterment* or improvement" (Dobzhansky et al. 1977, p. 509). Thus "progress may be defined as directional change toward the better" (Dobzhansky et al. 1977, p. 507). The difficult question to answer is: what is better? Evidently different criteria can be used for the definition of 'better' and hence different notions of progress will have to be distinguished. A large number of such criteria have been proposed by various authors (see, e.g., Huxley 1942, 1953; Rensch 1960; Williams 1966; Simpson 1967; Ayala 1974b). Thus, "progress" has a different meaning to these authors. Even the same author may use different concepts of progress. Simpson (1949) examined several criteria such as dominance, invasion of new environments, improvement in adaptation, increased specialization, control over the environment, increased structural complexity, and increase in the range and variety of adjustments to the environment. Some of these criteria may be applicable only to certain evolutionary trends, whereas others may be of more general use. Dobzhansky et al. (1977, p. 511) proposed as a possible general criterion "the increase in the amount of genetic information stored in the organism." They warn, however, that "one difficulty, insuperable at least for the present, is that there is no way in which the genetic information contained in the whole DNA of an organism can be measured. The amount of information is not simply related to the amount of DNA." Another criterion suggested by Ayala (1974b), which is of more limited applicability, is "the ability of organisms to obtain and process information of the environment" (Dobzhansky et al. 1977, p. 513). According to this criterion humankind is considered to be the most progressed species. It should be kept in mind, however, that this criterion could be divided into subcriteria and then humankind would not always come out on top of all species. It is well known that various animal species obtain information that is inaccessible to us (see the discussion of von Uexküll's *Umweltlehre* in Sect. 1.3.1).

The notion of the general and absolute superiority of the human species, which has been nourished by our Judeo-Christian tradition, is related to an uncritical use of the concept of progress (see also Stent 1978, p. 214). It is important to realize that general and absolute progress does not exist unless one is as presumptuous as to single out one criterion as the absolute and only true criterion. That is, however, not possible in an objective way. "No single criterion is a priori the best" (Dobzhansky et al. 1977, p. 513). Thus, different criteria will have to be applied to explore the different aspects of change.

In order to make it more explicit that progress is not a general and absolute notion but relative to the criterion used, it would be best to discontinue using the general term 'progress.' It is inaccurate and misleading to say and to believe that Western man has progressed over so-called primitive tribes. We have made technological progress and other kinds of progress. However, in other respects we have made no progress or have regressed. I have met many people in so-called underdeveloped countries who lead more harmonious lives than the average person in technologically advanced countries. In my opinion, many of those people who lead a simple life in harmony with nature have greater wisdom than addicts of technology who want to control nature.

I do not want to imply that all technology must be bad by necessity. I want do emphasize, however, that an equation of technological progress with the notion of a general and absolute progress is absurd. Fortunately, more and more people are recognizing this. So there may even be some hope that our so-called progress in Western and Westernized societies will not lead us into more and more misery and destruction.

The question of the general superiority of a certain race over others is equally misguided. One might select one criterion and one might find that according to that particular criterion certain individuals within one race have progressed more than others. This does not at all amount to any general superiority. For example, with regard to a particular intelligence test, one might find that samples of one race have a higher mean I.Q. than those of another race. What does that mean? It means that in one respect one sample of individuals scores higher than another. The matter is complicated by the fact that there are different ways of measuring intelligence (see, e.g., Gould 1981; Lewontin et al. 1984). Depending on which way is chosen, the result of the test may be different. Even if a general measure of intelligence were possible, this would assess only one quality, leaving out many others such as sensitivity, kindness, happiness, wisdom. Who could substantiate that intelligence (whatever it means) is more impo tant than kindness?

## 8.5 Modern Evolutionary Theory

### 8.5.1 Three Aspects of Evolutionary Theory

A detailed analysis of modern evolutionary theory is beyond the scope of this book. The reader is referred to Dobzhansky et al. (1977), Ho and Saunders (1979, 1984), Gould and Lewontin (1979), Mayr and Provine (1980), Gould (1980), Beatty (1980), Stebbins and Ayala (1981), Scudder and Reveal (1981), Good (1981), Tuomi (1981), Maynard Smith (1982), Hitching (1982), Wiley and Brooks (1982), Grene (1983), Grene and Burian (1983), Birx (1984), Brandon and Burian (1984), Pollard (1984), Sober (1984). However, since evolutionary theory usually is considered to be the most comprehensive biological theory, some of its philosophical bases and consequences shall be briefly discussed.

Three aspects can be distinguished in any evolutionary theory: (1) the general postulate that organic evolution has occurred, (2) the detailed reconstruction of evolution, i.e., phylogenetic reconstruction, and (3) a theoretical framework that provides an explanation of evolution, i.e., evolutionary processes. Often only the third aspect is referred to as evolutionary theory [see, e.g., Hull (1974, p. 50) who quoted Hempel (1965, p. 370)] stating that we must "distinguish what may be called the *story* of evolution from the *theory* of the underlying mechanisms of mutation and natural selection". Accordingly, "the claim that mammals arose in the Jurassic Period from ancestral reptiles is part of the story of evolution, a phylogenetic description, and not part of evolutionary theory. On the other hand, the claim that speciation can take place only in the presence of appropriate isolating mechanisms is part of one version of evolutionary theory. Phylogenetic descriptions are historical narratives; evolutionary theory is not" (Hull 1974, p. 50).

Much agreement exists with regard to the first aspect, i.e., the general postulate that organic evolution has occurred. As more and more fossils become known, the continuity of organic evolution becomes increasingly evident. There are, however, still many missing links and some critics have pointed to the resulting discontinuities in order to support versions of creationist alternatives. Evidence against these creationist alternatives is, however, very strong.

The second aspect, i.e., phylogenetic reconstruction, is much more controversial than the first one. Especially in groups of plants and animals with a poor fossil record, phylogenetic relationships are difficult to establish and therefore the phylogenetic reconstructions are often highly speculative.

With regard to the third aspect of evolutionary theory, controversy also continues: we find a great variety of more or less divergent views (see, e.g., Ho and Saunders 1979, 1984; Gould 1980; Stebbins and Ayala 1981; Waesberghe 1982; Wiley and Brooks 1982; Grene 1983; Goodwin et al. 1983; van der Hammen 1983; Pollard 1984; Dentor 1984). Nonetheless, a body of ideas, which is referred to as the Synthetic Theory of Evolution (STE), has become widely accepted and is still represented in most textbooks from the elementary to the most advanced level. In a response to increasing criticism it has, however, been more or less modified in different ways, and therefore it has become increasingly difficult, if not impossible, to characterize it. According to Wuketits (1978, p. 142) the following three main postulates are at the core of STE: (1) mutationism, (2) natural selection in the sense of external selection, (3) isolation. This implies that evolution results from natural selection of random mutations; it is the environment that selects (external selection), and speciation occurs through isolation mechanisms. Not all adherents of STE would completely agree with this formulation. Some of them would modify or qualify these generalizations; others would add additional postulates. As a result of this plasticity, it is difficult to criticize STE because one may be accused of having erected a "straw man" (see, e.g., Stebbins and Ayala 1981, p. 967). I shall therefore not try to criticize STE in general, but rather more specific principles or postulates that are characteristic of many, though not all, versions of STE.

## 8.5.2 Chance and Necessity

Monod (1970) presented STE in terms of two basic notions: chance and necessity. Thus, everything that has evolved, including humankind, is the result of chance and necessity. Chance occurs through mutations and recombination, which lead to novelty and diversity. Necessity applies to the selection of those variants that are best adapted. What is chance? According to one common definition, it is an event that occurs "at the intersection of two independent causal series" (Nagel 1961, p. 326). This definition obviously is based on the notion of linear causality. From the point of view of network thinking the really interesting and challenging question is the interconnection of the two so-called independent causal chains in the whole network. Is this interconnection due to very weak links or are there also at least some strong interactions? In the former case, it would be adequate to a certain extent to consider the two chains separately. Hence, the above definition of chance would have a good degree of adequacy. However, if it turned out that there are also strong interactions between the two so-called independent causal chains, it would be rather arbitrary to single out these two chains and to define chance in terms of them.

The consequences of these considerations for evolutionary theory are profound. If chance in evolution is not what we have been assuming it to be according to the above definition, then we shall have to change the theory of evolution. Thus, Ho and Saunders (1979, p. 573) go as far as to propose "that the intrinsic dynamical structure of the epigenetic system itself (i.e., the whole developing organism), in its interaction with the environment, is the source of nonrandom variations which *direct* evolutionary change."

Darwin had the admirable modesty to admit that the concept of chance represents only our ignorance: that which we do not understand we call chance. Modern biologists have not always taken Darwin as an example as far as the virtue of modesty is concerned. Students of biology are exposed to many dogmatic statements made in textbooks and publications of original research. For example, one can often read that evolution is the result of pure chance as far as the mutations are concerned with regard to adaption and survival. In this context chance is usually understood as the meeting of two totally independent causal chains. Monod (1970) also defines chance in that sense and comes to the conclusion that evolution is blind: "Pure chance, absolutely free but blind, at the very root of the stupendous edifice of evolution" (Monod 1970). I wonder whether evolution is blind or whether Monod was blind.

Instead of defining chance simply as the intersection of two independent causal chains and leaving it as that, I suggest at least the following modification: chance is the intersection of two causal chains whose interconnection in the network of the more inclusive system we do not yet know. I think that this definition has the advantage of being less dogmatic and final. Furthermore, it points to the problems that still have to be solved and does not obscure them.

According to our present scientific world picture events are integrated into the network of interactions of the whole system. To me 'interaction' means interdependence and even interpenetration (see Lewontin et al. 1984, p. 272). Interdependence can be described either in terms of determinism s. str. (simple actions and interactions), or determinism s. lato (including stochastic actions and interactions). If either simple or stochastic interactions will be demonstrated between the two so-called independent causal chains of evolutionary chance events, then evolutionary theory would have to be changed drastically because chance according to the above definition would no longer exist. Evolution would be the result of determination s. str. or s. lat. (see Chap. 6).

So far science. Now I shall try to go beyond the boundaries of science. If we shall find no evidence for the deterministic interpretation of what is now called chance, that does not necessarily imply a total lack of interdependence of the two causal chains. These chains (or rather the reality to which they refer) could be interdependent in a way that cannot be described through scientific laws, i.e., they could be interdependent in an irregular way. Irregular interdependence can be called chaos. Chaos need not lack harmony. In fact it can be harmonious in a way that is beyond conceptualization and lawful description.

In Taoism chaos is experienced as harmony (see Izutsu 1967). Harmony is the way of the whole (Tao). What we call chance or chaos may be activity within the harmonious whole. However, the harmony of the whole (Tao) cannot be described in terms of the scientific concepts of "chance" or "chaos." It is beyond conceptualization and therefore transcends not only the scientific concepts of "chance" and "chaos," but also that of "law." The chaos of the Tao is not the "chaos" of scientists, although there may be a relationship, and the laws that are referred to in Taoism or Buddhism (see, e.g., Da Liu 1974, p. 7) are not laws of science, which can be described symbolically, but they are intangible as it is pointed out for laws in the Prajna Paramita Hridaya sutra (unpublished translation by Master Lim):

> All the laws are intangible
> Neither manifested nor destroyed
> Neither unjust nor just
> Neither increase nor decrease
> In the center of the intangible nothing is seen or manifested.
> You do not have to think to understand it

Here we are beyond abstractions and thought and consequently beyond science. The Prajna Paramita Hridaya sutra "also referred to as the Heart sutra ... is to be grasped not through the intellect but with the heart – that is through one's own deepest intuitive experience" (Kapleau 1979, p. 179). Hence, "you do not have to *think* to understand it."

The terms law, lawfulness, order, harmony, and chance are used differently in different cultures and by different people in our highly pluralistic Western culture with all shades of the counterculture. Hence, apparently contradictory statements may have the same or similar meaning, whereas the same statement may have quite different meanings. For example the statement that there is order in the universe and its negation may have the same meaning depending on the definition of order. The same applies to lawfulness, harmony, and chance. To communicate well, one has to know how the terms are defined. And since the definition finally rests on primitive terms whose meaning is grasped intuitively one has to share the same intuition, i.e., state of consciousness. Thus, similar experience is important to grasp the significance of basic terms such as the above.

Returning to the scientific context of STE, I shall now briefly discuss Monod's second basic notion, namely necessity. Necessity operates through natural selection. Among the variants of organisms and populations that arise by chance, the best adapted are selected and in this way evolution occurs. Thus, natural selection is a central concept. I shall present two critical remarks on it, the first one on its extension and the second one on its intension.

(1) It has been demonstrated in certain specific cases that natural selection occurs. An example is the much quoted selection of the darkly colored moths in industrial areas where the habitat of the moths has become equally dark. What has not been demonstrated in my opinion is that natural selection is the sole or even the major moving force of evolution. This means that it is still debatable to what extent natural selection plays a role in evolution. In other words: the extension of the concept of natural selection is questionable. According to the neutrality theory of molecular evolution, much protein evolution does not result from natural selection (see, e.g., Kimura 1979). Similar claims have been made also at the macrolevel. Ho and Saunders (1979, p. 573) conclude that "the role of natural selection is itself limited: it cannot adequately explain the diversity of populations or species; nor can it account for the origin of new species or for major evolutionary change: a relative *lack* of natural selection may be the prerequisite for major evolutionary advance." Besides referring to a relative lack of natural selection, they also speak of "relaxed natural selection" [Ho and Saunders 1979, p. 586; see also Portmann (1965), Lewontin (1972), van Steenis (1980), Ho and Saunders (1984), and Pollard (1984) who, among others, have been critical of natural selection].

(2) The term and principle of natural selection have different connotations (intensions) (see, e.g., Bradie and Gromko 1981). One frequent implication is that selection is external selection, i.e., the environment of organisms selects the latter. According to this view, the organisms are passive as far as selection is concerned. However, this view has been challenged by proponents of "internal selection" (see Koestler 1967; Wuketits 1978; Riedl 1975, 1979). This latter concept refers to the selection of mutations by the organism: some of the mutations are accepted, others are rejected in the sense that the organism repairs the mutation and thus eliminates it. Strictly speaking, it is the organization of the organism that selects. Hence, organization and its development (epigenetics) becomes of paramount importance (see, e.g., Lovtrup 1981; Ho and Saunders 1984; Pollard 1984). Genes and their activity are thus seen in a more holistic context (see, e.g., Waesberghe 1982; Hitching 1982, p. 195).

Both the concept of internal selection and the epigenetic approach have far-reaching consequences. The organism is no longer seen as totally passive in evolution, but it is also active. It is selected, but it selects too. Concerning the biological and cultural evolution of humankind, this means that we are not the helpless victims of the environment; and hence there is no basis for a fatalistic view in this respect. Existential and social implications abound (see, e.g., Koestler 1967; Rose 1982; Rose 1983; Lewontin et al. 1984).

On the theoretical level the incorporation of the notion of internal selection into the framework of modern evolutionary theory leads to the systems theory of evolution (Wuketits 1978, 1983). As the name 'systems theory' indicates, it views organisms and environment as an inclusive system. This means that neither the organism nor the environment exists as separate entities as it has been tacitly assumed in Darwinism. Organism and environment form a whole because of a profound interpenetration (Lewontin et al. 1984, p. 272). Hence, the organism is not solely determined by the environment. Organism and environment co-determine each other and co-evolve as a result. The notion of circular causality and network thinking (see the chapter on causality) are appropriate in this context. Inasmuch as linear causality is implied in STE, the systems theory of evolution goes far beyond STE (Wuketits 1978, p. 146).

In concluding this brief sketch of a few selected issues, it seems no exaggeration to state that evolutionary thinking is in a ferment (Dentor 1984). Although the majority of biologists still accepts various versions of STE, there are more and more authors who question these versions (see, e.g., Ho and Saunders 1984; Pollard 1984). Much of the debate is due to a different emphasis of various phenomena. This may lead to a considerable modification of STE (see, e.g., Stebbins and Ayala 1981; Buss 1983; Takhtajan 1983). At which point one refers to a "new" theory appears to be somewhat arbitrary, especially in view of the protean plasticity of STE. On the other hand, certain alternative proposals, such as, for example, the nonequilibrium approach to evolution (Brooks and Wiley 1984), are a rather fundamental departure from much traditional evolutionary thinking. In view of all these developments, it is difficult to predict what kind of evolutionary theory will have emerged in 10, 20, or 50 years from now. I would not be surprised if that new theory would have only little in common with orthodox Darwinism.

## 8.5.3 The Status of Evolutionary Theory and Tautology

Questions have also been raised with regard to the status of STE in its various versions. Is it a predictive theory, or only a nonpredictive explanatory theory, or not even the latter?

Concerning the question of the predictive power of STE, many biologists admit that it does not allow for good predictions, if any predictions at all (see, e.g., Simpson 1964, p. 137). For example, "if we are to predict the course of [the evolutionary] development of a particular species, then we must be able to predict the successive environments to which the members of this species are to be exposed ... A river changes course and splits a single population into two, a new predator invades the territory of another species, a new mutant appears and must be accommodated or eliminated ..." (Hull 1974, p. 61). Can we ever hope to predict the evolution of species in the face of such complexity? It may be debatable whether probabilistic predictions may be possible under very specific conditions in certain restricted cases, but in general successful prediction seems unlikely. This does not necessarily mean, however, that other kinds of predictions are beyond reach. Hull (1974, p. 65) quoting Williams (1970, 1973) emphasized that we may have been looking for the inappropriate kind of predictions. Instead of trying to predict the future of "particular individuals, properties of individuals, or even taxa," we should rather predict "patterns of evolution" (Hull 1974, p. 65; see also Hull 1976, 1981). In this way we can be more successful according to Hull, and STE may then be considered to be a predictive theory. An example of a pattern prediction is the prediction of the superiority of the heterozygote (Hull 1974, p. 66). Ferguson (1976) and Williams (1982) list other examples and thus conclude that STE can predict. Williams (1982) points out, however, that the predictions of STE are not about the future, or in any case predictions about events at a specific future time do not play a central role in evolutionary biology. This is in contrast to physics where prediction of future events is common.

Although pattern prediction is more likely to be successful than the prediction of the evolution of taxa, problems exist also with regard to pattern prediction (see, e.g., Peters 1976, 1978). As a result, many biologists and biophilosophers deny that STE is a good predictive theory. A number of biologists such as, for example, Simpson (1964), rather consider it to be only an explanatory theory. Such a theory can explain events after they have occurred, but it cannot predict them. According to Simpson (1964, p. 147) this is characteristic of all historical sciences and their theories. It applies for example to certain psychological theories, such as Freudian theory or computational theories of the mind (Boden 1979, p. 116).

Some critics of STE go further, denying that it is a scientific theory. According to Popper (1974), STE is only a metaphysical research program. It does not qualify as a theory because in Popper's opinion it lacks both predictive power and testability. Bunge (1977), Tuomi and Haukioja (1979) and others pointed out, however, that combined with more specific hypotheses STE becomes testable. Popper recanted (see Riddiford and Penny 1984, p. 8).

Other critics of STE claim that it is basically tautological [see, e.g., MacBeth (1974), Peters (1976); for an exchange of views on this matter see

Peters (1978) and authors quoted by him]. The crux of the problem lies in the basic principle of natural selection or the "survival of the fittest." If "the fittest" are defined as those that survive, then "survival of the fettest" is reduced to the tautological statement of the survival of those that survive. The question is whether this tautology can be avoided. Many evolutionary biologists think so. Dobzhansky et al. (1977, p. 505), for example, propose that the following is a nontautological formulation of STE: "Among alternative genetic variants, some result in features that are useful to their carriers as adaptations to the environment. Individuals possessing useful adaptations are likely to leave, on the average, greater numbers of progeny than individuals lacking them (or having less useful adaptations). Therefore useful adaptations become established in populations." This formulation hinges on the notion of adaptation. The above authors admit that the concept of adaptation is difficult to define. Furthermore, it is not easy to apply it rigorously (see, e.g., Gould and Lewontin 1979).

Williams (1970, 1973, 1979) has pointed out that the problem of circularity and tautology results from a failure to acknowledge the necessity of primitive terms. Newton's first law is circular if force is defined by mass and mass is in turn defined by force. The circularity is avoided if mass is treated as a primitive term which is left undefined. Similarly, the tautology of the "survival of the fittest" is removed if "the fittest" is considered as a primitive term. Whether this strategy is useful and desirable is, of course, another matter.

Beatty (1980) discusses the semantic and the traditional views of theories and shows that the former is better suited for evolutionary theory. "On the semantic view a theory is not comprised of laws of nature. Rather, a theory is just the specification of a kind of system – more a definition than an empirical claim" (Beatty 1980, p. 410). Thus, "whether Mendel's law or the Hardy-Weinberg law is really a law of nature is irrelevant from the perspective of the semantic view of the synthetic theory" (Beatty 1980, p. 410). Empirical claims assert only that a particular breeding group is an instance of the definition of the law or theory. In other words: the crux is the domain of applicability of the theory or law (see Chap. 1.5).

## 8.6 Evolutionary Epistemology

The question of epistemology is whether we can acquire knowledge of the real world, and if so, how. (Naive) realists say yes, extreme idealists no. According to the latter our knowledge of the world is only an invention of our mind that we project into the world, often wrongly assuming that it coincides with the real world. Evolutionary epistemologists try to bridge the gap between the two extreme positions of realists and idealists. They postulate that we can know reality to some extent because there is a certain correspondence between our mind, i.e., our cognitive structures, and the structures of the real world. This correspondence is the result of adaptation to the real world due to natural selection. The adaptation "has been essential for the survival of man as soon as teleological action, i.e., conscious, goal-aimed, willed, and purposive action came into play. Thinking before acting requires right thinking" (Mohr 1977, p. 204).

According to Mohr (1977, p. 206) the term 'evolutionary epistemology' was first used by Campbell (1974) and Vollmer (1975). Many other authors such as Lorenz (1941, 1973) and Popper (1972, 1984) have greatly contributed to the evolution of this kind of epistemology. Recently, Riedl (1980) devoted a book entitled "Biologie der Erkenntnis" ("Biology of knowledge") to this topic. Von Ditfurth (1979) hailed Riedl's book as having the significance of a Copernican turn of thinking. In the following discussion of evolutionary epistemology, I shall refer mainly to Riedl's (1980) presentation (for other versions see, e.g., Wuketits 1984).

According to Riedl (p. 8), evolutionary epistemology solves the following age-old philosophical problems: (1) the a priori, i.e., the question why and how we consider certain ways of reasoning (such as reasoning in terms of causes, space, time, and purpose) as given independently of individual experience, (2) inductive generalization and how it relates to reality, (3) the quest for certainty, (4) sameness, (5) causality, (6) purposiveness, and (7) the foundation and validation of rationality. The solutions are the following: To 1: The a priori are a system of hypotheses that have evolved as a result of increasing adaptation to reality ["hypothetical realism" of Lorenz (1973)]. Hence, they are a posteriori in terms of evolution. They incorporate what organisms have learned during evolution and therefore fit the real world as the fin of a fish fits the water (p. 182). To 2: Reality shows a hierarchical order with high redundancy (i.e., lawfulness which implies repetition). Such order must exist because our ordered perception and thinking evolved as an adaptation to it. Our perception and thinking is thus a reflection of reality. It is ordered reality that has become conscious of itself. Induction is successful because of the existing order in the world and our adaptation to it. Thus, we generalize inductively in accordance with the laws or nature. To 3: Absolute certainty is, however, not attainable. Although we have the innate drive for it, we cannot reach it because of limitations of our senses and reason (p. 185). To 4, 5, and 6: Thinking in terms of sameness, linear causality and purposiveness is innate, and since it evolved also as an adaptation to reality, it corresponds partially with it. Scientific analysis has shown, however, that absolute sameness does not exist and that causes (efficient causes) and purposes (final causes) form a network (p. 183). To 7: There is no absolute beginning, nor are there first facts or final reasons or proof. Hence, rationality cannot be founded in absolute terms. It must be seen as the result of increasing adaptation to the order of reality. If it is applied in a context in which it has not yet been tested and selected during evolution, it can fail us miserably and lead to devastating consequences. Therefore, the limits of rationality must be clearly seen.

Riedl also deals with the consequences of his solutions. He points out that many dualisms are overcome such as materialism/idealism, rationalism/ empiricism, determinism/indeterminism; in other words: the dualisms between matter and spirit (body and mind), reason and experience, determination and uncertainty. All this is achieved according to Riedl simply by basing epistemology on evolutionary theory and by drawing the consequences.

What are Riedl's basic assumptions and postulates?

(1) The methodology of natural science permits the solution of the epistemological problems. Particularly, objective testability is important (p. 11).

(2) Evolutionary theory, which is basic to epistemology, borders at certainty, i.e., the probability that evolutionary theory is correct approaches 100% (p. 12).

(3) Adaptation of organisms always entails an increase in their information of the environment significant to them (p. 7). Hence, evolution, as it leads to increasing adaptation, is a process of increasing cognition. "Leben selbst ist ein erkenntnisgewinnender Prozeß" (Life itself is a process of acquiring knowledge) [Lorenz (1971, p. 231), quoted by Riedl (1980, p. 12), (1984, p. 1)].

(4) Life is goal-oriented and success-oriented. "Das Lebendige jedoch bedarf fortgesetzt auch des Erfolges" (The living, however, is continually in need of success) (Riedl 1980).

(5) As a corollary of 3 and 4, modes of perception and thinking have been selected that come closest to a reproduction of reality.

(6) Consequently, there is a certain correspondence of the order of perception and thinking on the one hand and the hierarchical order of nature on the other hand. In other words: there is at least a partial isomorphism between patterns of nature and patterns of perception and thinking.

(7) The foundations of reason are innate, i.e., hereditary. Thus, the urge to attain truth (certainty) and thinking in terms of sameness, linear causality, and purposiveness are not acquired, but inherited. The same is true for our expectation of the constancy of laws of nature (p. 68).

To these seven basic assumptions and postulates I want to offer the following criticism:

To 1. As I have been trying to point out throughout this book, scientific methodology is limited for a number of reasons. How then can scientific methodology alone provide a sufficient basis for an understanding of reality?

To 2. As I have tried to indicate briefly in the preceding section of this chapter, modern evolutionary theory is still controversial in many respects. Inasmuch as aspects of this theory are shaky, evolutionary epistemology is debatable since it is based on evolutionary theory.

To 3. As I mentioned already, adaptation is not always easily demonstrated or even defined. Furthermore, it is not evident that adaptation *always* entails an increase in information. The harmonious integration of an organism in an ecosystem need not imply the notion of information or cognition. In fact, the latter notions appear to be more restricted in their application.

To 4. As I tried to show in the chapter on teleology, the notion of goal-orientation is also too limiting for a full understanding of life. The same applies to the idea of success-orientation, which seems to be fed to a great extent by capitalist society.

To 5. On what basis can one say that our mode of thinking, which is embedded in Western tradition, has been selected and therefore constitutes an adaptation? It may be an adaptation to some extent. But in general it is questionable whether we survive because of our modes of thinking or in spite of them. So-called primitive tribes rely much more on intuition than on reason. Our intellectual culture of Western and Westernized society has not yet passed the test of survival. We might be close to extinction due to an exaggerated emphasis of the intellect. This would not seem to indicate that the intellect mirrors the order of reality.

To 6. Projecting our notion of order into reality may give the impression of correspondence, which may indeed exist to a certain extent. However, to tackle the question of order in nature requires an expanded consciousness. A Western bias that pays only selective attention to a certain aspect of nature appears too restrictive to cope with the question of order in nature. (Concerning hierarchical order see Chap. 4.)

To 7. The dichotomy between innate and acquired features is not very useful because any feature results from an interaction of genetic and environmental factors (see, e.g., Weiss 1973; Lewontin et al. 1984). However, there are traits that appear rather stable in different environments and are thus called innate in a simplistic manner. Other traits that are easily influenced by the environment are usually classified as "acquired." I think that much of our reasoning including our categories of reasoning is influenced by our environment, i.e., our upbringing in a traditional way of thinking. Thus, our thinking in terms of sameness, linear causality, purposiveness, and our urge for certainty may be largely the result of conditioning. If it were innate how would it be possible that certain liberated individuals are able to transcend this kind of reasoning completely? And how could we explain that in some other cultures people care less about certainty but enjoy life as it comes? It seems that our potential is underestimated by Riedl. We are not as rigidly programmed as Riedl assumes. We are extremely plastic and we can develop and unfold far beyond the ordinary imagination (see, e.g., Bohm 1973, 1980; Lewontin et al. 1984; Hayward 1984; Krishnamurti 1985; Krishnamurti and Bohm 1985).

In spite of this criticism, I think that evolutionary epistemology has great merits. It is dynamic to a certain extent and has dealt with a number of sterile dualisms, such as those of idealism versus materialism, rationalism versus empiricism, determinism versus indeterminism, a priori versus a posteriori. In general it has broken down the barrier between our biological and spiritual aspects and is thus helping to bridge the gap between the biological sciences and the humanities. If it could overcome other dualisms, such as innate versus learned, if it could incorporate a broader base of more dynamic evolutionary thinking, and if it could avoid a narrow Western bias, it might even lead to a new Copernican turn.

### 8.7 Evolutionary Theory and Human Values

Evolutionary theory plays a central role in biology and therefore is related to many biological disciplines ranging from molecular biology to sociobiology. In addition, it may influence in many ways aspects of our life including our values and ethics (see, e.g., Kozlovsky 1974; Lumsden and Wilson 1981, 1983; Lewontin et al. 1984). In a discussion of values it is important to realize that there is a feedback between evolutionary theory and values:

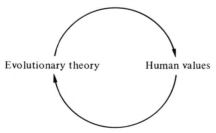

Hence, evolutionary theory does not only influence our values but it is itself influenced by our values as it is typical of scientific methodology (see Sect. 1.3.4). With regard to Darwinism, values (ideology) of nineteenth century England have had a great impact. Radl (1930, p. 18) wrote: "Darwin merely transferred the prevailing English political ideas and applied them to nature." Thus, he projected into living nature "his English society with its division of labor, competition, opening up of new markets, 'invention', and the Malthusian 'struggle for existence.' It is Hobbes' *bellum omnium contra omnes'* (war of all against all)" [letter by Karl Marx to Engels (1862), quoted by Gould and Eldredge (1977, p. 145)]. Even the idea of gradualism (gradual evolution) "was part of the cultural context [of the laissez-faire liberalism of Victorian England]" (Gould and Eldredge 1977, p. 145). Once the projection of these values had occurred and nature was seen as competitive and full of struggle for existence, egotism and competition in human society could be sanctioned as natural[1]. In this vicious circle values tend to be conserved. It is therefore not surprising that in modern capitalist society we still find the same or similar values providing fertile ground for neo-Darwinism (STE). Thus, in terms of modern capitalist thinking, evolution is expected "to operate as a manager of a multinational enterprise would direct his business: parsimonously pursuing his strategy, cost-efficient (Solbrig 1979), and with an optimal outcome (Horn 1979)" [Poppendieck (1981, p. 208) who is critical of this approach and welcomes diverging opinions]. Neo-Darwinism, reinforced by capitalist ideology, in turn is used to "justifiy" the status quo that nourishes it. For example, modern sociobiology (Wilson 1975; Lumsden and Wilson 1981, 1983) that is based on neo-Darwinistic thinking and its values has been used and misused in this respect (see, e.g., Sahlins 1977; Montagu 1980; Lewontin 1981; Lewontin et al. 1984). Since it is a synthesis of different biological disciplines, it reflects additional values that are also predominantly conservative such as a reductionistic, mechanistic bias that may be traced back to the Cartesian view of organisms as machines (Lewontin 1981; Lewontin et al. 1984; Levins and Lewontin 1985) Furthermore, economic concepts, such as cost-benefit analysis of our modern capitalist societies, have been incorporated (Lewontin et al. 1984, p. 59). Many values and postulates that play an important role in much sociobiological thinking are well captured in "The Gospel according to Sociobiology" (Boucher 1981)[2]:

*Synthesis*

In the beginning the Lord Gene created the heavens and the earth, and all the beasts thereof, and even the Groves of Academe. And many were those who dwelt in the Groves, and eminent their degrees, and numerous their words, even till the journals did overflow, so that they would not perish.

---

[1] Silvertown (1984, p. 185) gives the following quotation by John D. Rockefeller (ca. 1900): "The growth of a large business is merely a survival of the fittest ... This is not an evil tendency in business. It is merely the working-out of a law of nature and a law of God." More recently, Richard Nixon, the former president of the United States reaffirmed that "America's competitive spirit, the work ethic of this people, is alive and well" (Psychology Today, Sept. 1972, p. 53).

[2] Reprinted from "Perspectives in Biology and Medicine", Vol. 25, pp. 63–65, by permission of the author and The University of Chicago Press.

But the dwellers of Academe became clamorous, and frictions did prevail, and great was the heat thereof. And the dwellers began to speak in many tongues, and strange were their words, each unto the other. Some spake in Anthropology, and some in Psychology, and some in Economics, and some in Sociology, and some in Music, and some in History, and some even in Literature. And the people of the Lord Gene spake in Biology, but none would heed them.

So the Lord spake unto the people of Biology, saying: "Unto thee shall be given a New Synthesis, that shall be a mighty Synthesis, that shall reconcile all the peoples of Academe, the ones unto the others. And no longer shall they speak in Anthropology, nor in Psychology, nor in Economics, nor in Sociology, nor in Music, nor in History, nor even in Literature; but they all shall speak in Biology, every one. And their tongue shall be Biology no longer, but they shall be born again, and their words resynthesized, and their tongue shall be called SOCIOBIOLOGY. The people shall know their true natures, and they shall all behave, each according to his nature, and all shall dwell in peace and harmony. Great shall be their eminence, and many their disciplines, and weighty their publications, and neither shall they perish, but their fitness shall be maximal forever." Thus sayeth the Lord.

The commandments of the Lord Gene unto the people of SOCIOBIOLOGY:

I am the Lord thy Gene, which brought thee to life out of the slime of the ocean primeval.

Thou shalt have no other gods before me. Thou shalt not bow down to church, nor state, nor love, nor money, nor the needs of the fellow creatures, but shall act only so as to Increase thy Fitness. For I the Lord thy Gene am a jealous god, visiting the maladaptiveness of the fathers upon the third and fourth generations of those that do not look out for their Fitness.

Thou shalt not take the name of the Lord thy Gene in vain; but shall serve it with all thy heart and soul and mind and body and phenotype.

Honor not thy Father nor thy Mother, for their interests and thine are in Conflict. Neither honor thy sister nor thy brother, except according to their Coefficient of Relatedness unto thee.

Thou shalt commit adultery whenever possible, that thine Offspring be multiplied, but the cost of Parental Care be borne by others.

Likewise shalt thou bear false witness, and Cheat, and Fail to Reciprocate Altruism, but with a guise of Sincerity, that others might not know thy Cheating.

Especially shalt thou covet thy neighbor's wife, and house, and field, and ox, and ass, and everything else that is his; for with these thou canst Reproduce More, and he less, and thy Fitness shall Increase over his.

Thus sayeth the Lord.

*The Sermon on the Mount*

And seeing the multitudes, he went up unto a mountain, and his graduate students came unto him. And he opened his mouth, and lectured, saying.

Blessed are the Strong in Will, for theirs is the alpha position, and the right to many mates.

Blessed are the Rich, for they shall inherit the goods, and survive well, and reproduce greatly.

Blessed are they who neither Hunger nor Thirst, for they shall have time to compete for mates.

For I say unto you, that except your Fitness shall exceed the Fitness of the Scribes and the Pharisees, thy genes shalt not increase in future generations.

You have heard it said in olden times; thou shalt commit Adultery.

But I say unto you, that whosoever doth not always look upon a woman with Reproduction in his Mind, shall fall behind in the Race of Fitness.

You have heard it said, An eye for an eye, and a tooth for a tooth. But I say unto you, An Altruism for an Altruism, unless thou canst Cheat and Not Reciprocate.

Be ye therefore Adaptive, even as your Theory, which is in SOCIOBIOLOGY, is Adaptive.

After this manner pray ye: Our Genotype, which art in Phenotype, Maximized be thy Fitness. Thy kin select, thy race protect, on earth as it is in theory. Give us this day our Parental Investment, and forgive us our Altruism, as we deceive those who are Altruistic to us. And lead us not into Maladaptiveness, but deliver us from Unfitness. For Thine is the Behavior, and the Selection, and the Reproduction, for ever and ever. Amen.

There have always been individuals with different views based on different values (e.g., Kropotkin 1902). Especially since the 1960's alternatives have been proposed more frequently. Thus, for example, the importance of mutualism and cooperation has been underlined at all levels of organization by an increasing number of authors (see, e.g., Lewis 1973; Novak 1975, 1982; Haken 1977; Richmond and Smith 1979; De Robertis and De Robertis 1980; Merchant 1980; Margulis 1981; Margulis and Sargan 1985; Axelrod and Hamilton 1981; Boucher et al. 1982; Axelrod 1984; Boucher 1985). May (1982) estimated "that empirical and theoretical studies of mutualistic associations are likely to be one of the growth industries of the 1980's." To what extent these studies will affect our values remains to be seen. So far our "theories of mutualism are still basically mechanistic, mathematical, fitness-maximizing, and individualistic" (Boucher 1985). There are, however, attempts to go beyond these presuppositions. For example, Lewontin et al. (1984) criticize dualistic thinking that is deeply rooted in Western culture. They underline that in nature there is no opposition of organism and environment. Organism and environment form a whole. If we are totally aware of this wholeness, alienation cannot occur because there is nothing to be alienated since the individual does not exist as such. This means that organisms can no longer be looked upon as respondants to the environment: "Organisms do not simply adapt to previously existing, autonomous environments; they create, destroy, modify, and internally transform aspects of the external world by their own life activities to make this environment" (Lewontin et al. 1984, p. 273) (see also Chaps. 9, 10).

This interpretation of organism and environment and related holistic views expressed by an increasing number of biologists and philosophers, based on values that differ rather fundamentally from those prevalent in Victorian England and modern capitalist societies, does not only affect evolutionary thinking, but the whole of biology, other sciences such as psychology, and society. The consequences are profound and have many ramifications. A more general awareness of these implications could amount to a fundamental transformation of science and society (see, e.g., Leiss 1972; Bateson 1972, 1979; Novak 1975, 1982; Ferguson 1980; Merchant 1980;

Birch and Cobb 1981; Skolimowski 1981; Rose 1982; Capra 1982; Rose 1983; Hayward 1984; Levins and Lewontin 1985)[3]. It could lead to more peace, harmony, and understanding.

Many of the alternative values that have become more frequently expressed since the 1960's are part of or related to Maruyama's (1974) and Johnson's (1977) mutual causal paradigm (see Sect. 1.4, Table 1.1) [see also Skolimowski's (1981) "eco-philosophy"]. The way of thinking associated with these values is often pluralistic: it accepts contrasting (even contradictory) models, theories, descriptions, viewpoints, values, etc. as different perspectives of nature (see, e.g., Varela 1979; Mishler and Donoghue 1982; Lewin 1983; Ames 1983; Lewontin et al. 1984; Rutishauser and Sattler 1985). It recognizes that any value or philosophy imposes a constraint and thus limits our understanding. Hence, a plurality of values and philosophies may be enriching and liberating (see, e.g., Gould and Eldredge 1977, p. 146; Rose 1982; Rose 1983).

Post-structuralists such as Derrida and Culler have emphasized this point in terms of "deconstruction" (Culler 1982). Any philosophy needs deconstruction which "is to show how it undermines the philosophy it asserts" (Culler 1982, p. 86). The same applies to values, theories, and other kinds of postulates including interpretations of poetry. If we fail to deconstruct, our views may become too rigid and one-sided and our science, philosophy, and life too impoverished.

## 8.8 Summary

*Introduction.* There are a number of typical kinds of questions that are asked in scientific research. Simpson distinguishes the following: What?, How?, What for?, How come? The last question is characteristic of evolutionary biology, although the others play also a role in this discipline.

*Universality of change.* Change is universal. The impression of stability or permanence appears to be due to very slow change in contrast to quick catastrophic changes. Thom's catastrophe theory can cope with slow and gradual changes as well as quick catastrophic changes and the whole continuum between these extremes. However, it is based on the permanent forms of seven elementary catastrophes. Thus, the question arises whether permanence is only a tool of reference-points to describe change or whether it is real. Usually, the laws according to which change occurs are considered to

---

[3] See also the excellent CBC series "A Planet for the Taking" by David Suzuki (CBC, Toronto).

be permanent. However, a number of scientists and philosophers such as Woodger and Bergson have questioned the validity of this assumption. Empirical evidence appears to be insufficient to decide whether nature is lawful in the sense of unchanging permanent laws. In the face of this un-decidability the task of the scientist is to adapt his models and theories as much as possible to the fluidity of nature. It is pointed out that modern evolutionary theory still retains static elements. Hence, a major challenge of evolutionary theory is to become more dynamic and more evolutionary.

*Historicity.* Science that deals with historical aspects is historical science. Major problems of historical science such as prediction, explanation, and uniqueness are discussed. Is the limited or lacking predictive power of histori-cal theories due to insufficient knowledge of parameters or an impossibility in principle? Or have we failed in our predictions because of focusing on the wrong kinds of predictions? Concerning explanation the question is whether it follows the covering-law model. If it did, laws would be required. Since it is difficult to find such laws, it has been claimed that historical science can provide only narrative explanations, which do not depend on laws but only on singular antecedent events. The uniqueness of history and historical events has been thought to be the reason for the lack of laws in history and hence for the necessity of narrative explanations. However, two levels of unique-ness have to be distinguished: the conceptual and the experiential (non-conceptual) levels. At the conceptual level unique events are described in terms of general concepts. At the experiential level events are totally unique and therefore cannot be analyzed scientifically. Any attempt at a scientific analysis loses the fullness, richness, wholeness, and holiness of the unique event. The emphasis of uniqueness varies between cultures such as, for ex-ample, Western and Chinese cultures; a considerable divergence in the life-styles and cultural evolution may be associated with this difference.

*Evolution and progress.* Evolution may be defined as change, or as direction-al change that leads to irreversibility, or as progressive change. The latter definition is too narrow. General progress cannot be ascribed to evolution. Progress occurs only with regard to certain features, i.e., progress is only progress in terms of a certain criterion. Since progress in general does not occur, general superiority cannot result. Hence, one can neither claim that human beings in general are superior to animals, nor that any human race is superior to others.

*Modern evolutionary theory.* Three aspects can be distinguished with regard to evolutionary theory: (1) the postulate that evolution has occurred, (2) the detailed phylogenetic reconstruction, i.e., the story of evolution, and (3) the

208

explanation of the evolutionary processes by a theory (or system of theories) that has been called evolutionary theory in the narrow sense. Many controversial issues still exist, especially with regard to phylogenetic reconstruction and evolutionary theory s. str. The most commonly accepted version of the latter is called the Synthetic Theory of Evolution (STE). Even within STE a variety of different versions exist and thus the exact formulation of STE is not an easy task. Two fundamental postulates of various versions of STE are critically discussed under the heading of chance and necessity. The common definition of chance as an event at the intersection of two independent causal chains is not satisfactory since it does not pose the crucial question of how these causal chains are interconnected in the total network of the most inclusive system. If a lawful interconnection were found between the two causal chains, a rather fundamental change of STE would be required. If no evidence for such regularities turns up, that does not mean that the two causal chains are totally independent. They could be interdependent in an irregular yet harmonious way. Harmony and chance need not be mutually exclusive. Definition and underlying experiences of lawfulness, order, harmony, chaos, and chance may differ considerably. The second major component of STE, namely necessity, operates through natural selection. A few critical remarks on natural selection are made. It is questioned whether natural selection plays an all-important role with regard to evolution. Furthermore, it is emphasized that the notion of external selection (by the environment) has to be complemented by the concept of internal selection. The latter focusses on organization and development of the whole organism (epigenetics) in contrast to an overemphasis of genes and gene action. From this point of view the organism is not only selected, but selects also. This codetermination of organism and environment, which leads to co-evolution of the two, is stressed by the systems theory of evolution. Finally, it is briefly discussed whether STE is predictive, or only explanatory, or tautological. The semantic view of theories is applied to STE.

*Evolutionary epistemology.* Evolutionary epistemology, which bridges the gap between naive realism and extreme idealism, postulates a partial correspondence between our mind, i.e., our cognitive structures, and the structures of the real world due to natural selection and increasing adaptation of our perception and thinking to reality. Riedl's (1980) exposition of evolutionary epistemology, which has been hailed as having the significance of a Copernican turn, is discussed with regard to his proposed solution of age-old philosophical problems and their implicit assumptions and postulates are evaluated.

*Evolutionary theory and human values.* Because of a feedback between evolutionary theory and human values, evolutionary theory does not only influence human values but is itself influenced by our values. Thus values of Victorian England and modern capitalist societies have been projected into Darwinism in its traditional and modern form (STE). And Darwinism in turn reinforces values such as competition in capitalist societies. This feedback favors strongly the retention of the status quo. Sociobiology is one of its manifestations. Nonetheless, alternatives have been explored, especially since the 1960's. Holistic approaches that emphasize cooperation and integration have been adopted by at least some biologists and the beneficial conse-quences for society have been underlined. Furthermore, there seems to be an increasing awareness that pluralism in science and society can be enrich-ing and liberating. Post-structuralists have emphasized this point in terms of "deconstruction."

# 9 What is Life?

"Living organisms can be understood only when they are
considered as part of the system within which they function" (Dubos 1981, p. 37)
"Present attempts to develop adequate principles of life
represent perhaps the greatest conceptual crisis in the history of science" (Davenport 1979, p. 2)

## 9.1 Introduction

The nature of life has been much discussed by philosophers as well as scientists
(see, e.g., Schrödinger 1944; Bertalanffy 1952, 1975; Portmann 1960, 1974;
Waddington 1961, 1968-72; Grene 1965, 1974; Jonas 1966; Blandino 1969;
Jacob 1970; Black 1972; Jeuken 1975; Canguilhem 1975; Elsasser 1975,
1981; Grene and Mendelsohn 1976; Heidcamp 1978; Atlan 1979; Bateson
1979; Buckley and Peat 1979; Varela 1979; Morin 1980; Crick 1981; Mercer
1981). Quite often it has been implied that a characterization or definition
of life must consist of a list of properties (or at least a single property) that
are exhibited by life, but are absent in nonliving nature. Such an approach is
based on the belief in a dualism between life and inanimate nature. During
the last decades, systems thinking has focused attention on more inclusive
systems, such as ecosystems, that comprise living organisms as well as so-
called nonliving matter. A characterization of an ecosystem need not imply
an absolute dualism between life and inanimate nature, but may instead
emphasize the characteristics of the whole ecosystem. From this vantage
point, life and reality may be seen in a more global perspective.

Evidently, the meaning of the question 'what is life?' depends on prior
philosophical assumptions. To an adherent of essentialism it means seeking
the essence of life. To someone who does not subscribe to essentialism but
who nonetheless presupposes a dualism between living and nonliving nature,
the question may concern the search for a property or properties that are
necessary and sufficient to define life. On the other hand, to a reductionist
the question is not so much the distinction of a property or properties that
are unique to life, but rather their reduction to the level of nonliving matter.
An integrated view of whole ecosystems may again imply different philosophical asumptions (see, e.g., Lewontin et al. 1984, Chap. 10). In general
terms, the meaning of the question 'what is life?' derives from the meaning

of "what is" and the extension of the concept of "life," i.e., whether "life" refers only to living organisms or to the whole biosphere.

## 9.2 Vitalism

According to the most encompassing form of vitalism, the whole cosmos is animated (animism). However, a more common form of vitalism, which has been of concern to some biologists, implies that life is fundamentally different from inanimate matter because it is endowed with a vital principle in the form of a vital substance, a vital fluid, or a vital force to which different names have been given such as entelechy, élan vital, soul, etc. (see, e.g., Blandino 1969). The assumption of such a vital principle answers the question 'what is life?' or 'what is a living organism?' at once. A living organism is an object that is infused with the vital principle. The orderly development and functioning of the organism occurs because the organism is directed by this principle.

Vitalism dates back to antiquity and before. Vitalistic thinkers of the nineteenth and especially the twentieth century are usually called neovitalists. One well-known neovitalist was the German zoologist Hans Driesch (1867–1941). He was a student of the materialist biologist Ernst Haeckel under whose influence he published an analytical theory of organic development (1894). Although this early work was written in the spirit of a machine theory of life, it shows already finalistic tendencies. His later work in experimental embryology, much of which he carried out with sea urchins at the famous zoological station in Naples (Italy), led him to postulate a vital force that directs the orderly development of organisms and in that sense constitutes a teleological force. He termed that force 'entelechy,' a name already used by Aristotle. The literal meaning of 'entelechy' is 'having its goal in itself.'

Driesch's vitalism is elaborated in his two-volume work on the "science and philosophy of the organism" (1908) and several other books in German. The empirical foundation for his conclusions are summarized in many embryology texts [see, e.g., Browder (1980) and by Blandino (1969)]. One of his famous experiments was the following. He divided a very young germ of a sea urchin into two halves and observed that each half formed a complete organism. If a machine is cut into two halves, he argued, it stops functioning. Hence, there is a fundamental difference between an organism and a machine. An organism "knows" its goal due to its entelechy, which directs its development and proper functioning even after severe disturbance and injury.

Many kinds of criticism have been leveled against vitalism and neovitalism. I shall present four principal ones:

(1) It has been said that the postulate of a vital principle "such as an entelechy is too successful, too general, and gives us no light upon the particular case" (Woodger 1967, p. 266). This means that a vitalistic doctrine is no scientific theory. A scientific theory by definition must be testable, i.e., it must allow us to derive particular test implications. Vitalistic and neo-vitalistic doctrines, it is said, fail on that account. They do "not indicate under what circumstances entelechies will go into action and, specifically, in what way they will direct biological processes: no particular aspect of embryonic development, for example, can be inferred from the doctrine, nor does it enable us to predict what biological responses will occur under specific experimental conditions" (Hempel 1966, p. 72). According to this criticism vitalistic doctrines are rejected as unscientific, but they are not criticized because they imply a nonmaterial agent. In fact, Hempel (1966, p. 72) emphasizes that Newtonian theory also invokes a nonmaterial agent in the form of gravitational forces. However, the latter theory is testable, and has explanatory and predictive power.

Although the above criticism may apply to many forms of vitalism, I am not convinced that it is generally valid. Instead of rejecting all forms of vitalism categorically as unscientific, I think we should remain open-minded whether at least some forms of vitalism may be able to provide certain kinds of explanations and/or predictions. For example, certain forms of Taoist health techniques as mentioned by Mencius (see, e.g., Da Liu 1974, p. 14) invoke a vital force called ch'i. Activation and transformation of ch'i through specific breathing methods may lead to predictable results. Thus, one can, for example, predict that the activation of ch'i will enable two men or women to lift a heavy person above the ground just with their little finger. I have witnessed such events to my great amazement. One can even make a controlled experiment. If ch'i is not activated the same people are unable to lift the person.

In conclusion, I want to make two observations with regard to the criticism that vitalism is unscientific. First, at least some forms of vitalism may have some explanatory and/or predictive power and in that sense are not necessarily unscientific. Further investigation along these lines might be profitable in several respects. Second, not all forms of vitalism are intended to be scientific. Criticizing such doctrines as being unscientific is beside the point unless one claims that anything unscientific is meaningless. Probably Henri Bergson (1859-1941) could be mentioned as a representative of an unscientific form of vitalism. Bergson (1889, 1907) called the vital principle 'élan vital.' It is experienced intuitively. Discursive thought alone cannot grasp it. Hence, rational criticism of Bergson's vitalism is misguided. The élan vital as it flows and creates spontaneously must be felt and experienced.

Bergson also wrote a book on laughter (*Le rire*, 1900). This again demonstrates how much his attitude toward life differs from that of analytic philosophers. To the latter, laughter does not further our understanding of life. To Bergson laughter may be a profound communion with life. Hence laughter may bring us closer to the mystery of life than discursive thought, and thus it may create harmony and happiness. If you do not agree with this, go to Thailand, "the land of smiles," or other countries where people smile and laugh a lot, and you will be amazed about the happiness and wisdom of many of those people. If you cannot visit the "land of smiles" or if the Thais will have become too westernized by the time you will arrive there, just seek the company of people who laugh and make you laugh and let yourself be influenced. Or laugh madly for five or ten minutes. This might do you more good and solve more of your problems and worries than much thinking (see also Cousins 1979). I do not want to suggest that laughter is the answer to all problems, nor do I want to imply that thinking is useless. But I want to convey that laughter, spontaneity and direct experience can open up strata of life that cannot be reached by the intellect.

(2) Returning to scientific aspects of vitalism, a second objection to it concerns the nature of the vital principle. It has been argued that such a principle cannot be demonstrated empirically as something that occurs over and above atoms and molecules. Furthermore, it is difficult to imagine how such an additional principle could have arisen during evolution.

This criticism again need not apply to all forms of vitalism. As Hull (1974, p. 129) points out, one trend in modern science has been "the shifting of key scientific concepts from the category of things and substances to the category of properties, especially relational and organizational properties. Life is no more a thing than is time, space, gravity, or magnetism. One might well add *mind* to this list." If vital forces are also viewed in this way, then "vitalism becomes a coherent position and the major objections to it can be circumvented" (Hull 1974, p. 129). To explain magnetism "we need not postulate a magnetic heaven as a final resting place for good magnetic fields" (Hull 1974, p. 129). The same may apply to vital forces and if we conceive neovitalism in this way it may come close to or merge with the generally accepted way of modern biology. Meyen (personal communication) has pointed out to me that Driesch's notion of entelechy is in fact not so different from modern notions of self-regulation of living systems.

I mention all this to demonstrate that vitalistic doctrines are extremely diverse ranging from nonscientific experiences to views compatible with scientific theories or doctrines. Hence, to label a biologist as a neovitalist does not mean very much except that it is derogatory and condescending from the point of view of those who think they are above all that but very often are simply ignorant of the many forms of biological approaches that

215

at times are classified as neovitalism. Instead of making blanket condemnations of neovitalists it would be more appropriate and more profitable to analyze particular approaches and theories on their own merits. Vitalism has become a bad and repugnant word for most biologists. Unfortunately, many are already biased when they hear or read the word and thus deprive themselves of a more positive evaluation.

(3) A third major criticism of vitalism is pragmatic. It is said that vitalism has not been successful as a scientific theory, whereas its rival doctrine, i.e., mechanism, has been crowned by success. Even where mechanism has failed, it has been considered to be at least of heuristic value, i.e., it may aid to find better models and theories. Vitalism, on the other hand, is accused of lacking heuristic value. Extreme critics claim that it is of no use at all.

In response to this total rejection of all forms of vitalism and the acceptance of mechanism on pragmatic and heuristic grounds, Woodger (1967, p. 269) pointed out that "heuristic success as such is no guarantee of truth, neither is it any guarantee that a given method will go on being successful indefinitely. Still less does it provide a *reason* for completely neglecting the study of other possibilities." One may also question whether vitalism is completely devoid of heuristic value. Woodger (1967, p. 266) reminds us that "vitalistic writings have their valuable side. They represent an adventure in thought – the exploration of a possibility ..." Vitalistic doctrines may "have the merit of calling attention to aspects of the organism which any adequate theoretical biology will have to take into account" (Woodger 1967, p. 267) (see also Canguilhem 1975).

(4) Vitalism can be accused of being dogmatic whenever it claims to possess the ultimate answer to the question of what is life. In this sense vitalism blocks progress in science. The mystery of life is "explained" by the vital principle and thus the essence of life is supposedly grasped. Nothing else but uninteresting details are left to be found out. Such dogmatism is not only short-sighted, but also destructive inasmuch as it leads to intolerance. Factual details are completely devaluated because they have no more any bearing on the truth, which supposedly is already established beyond doubt.

It is possible that certain vitalists are guilty of such hubris and it is even more likely that many have such tendencies. Unfortunately, the same can be said of mechanists (see, e.g., Crick 1966, 1981; or Monod 1970). If we really want to further understanding we are well advised by Woodger (1967) who stressed the need of intellectual humility, more cooperation and less competition between adherents to rival theories. "The notion that one theory excludes another arises from our desire to regard our work as conclusive and exhaustive of what there is to know. But the simple consideration that one system of abstraction cannot possibly be exhaustive will show us how mistaken such an attitude is" (Woodger 1967, p. 271).

216

## 9.3 Mechanism

Some people claim that many women suffer from penis envy. I do not know whether this is true, and if so, to what extent. I know, however, that many biologists suffer from physics envy. They look upon physics as the most exact and most advanced science with the most general theories that have great explanatory and predictive power. Furthermore, they take physics as the model science and the standard of any other science. In this perspective biology does not fare well. It has few general laws and theories and its most comprehensive theory, which is evolutionary theory, has little if any predictive power. Hence, in comparison with physics, biology is still at a pre-Newtonian stage. It has not even a body of theory comparable to classical mechanics as far as generality and predictive power is concerned [for a more detailed discussion of the relation between biology and physics see, e.g., Simon (1971, chapter 1); Bertalanffy (1975); Ruse (1977); Elsasser (1975, 1981)].

For a long time the exactness, simplicity, generality, and predictiveness of classical mechanics have been an ideal for many biologists. Consequently, they have tried to model biological theory after it, i.e., to explain life in terms of classical mechanics. Thus, living systems are simply viewed as matter in motion obeying the laws of classical mechanics. Critics who challenged the foundation of this kind of mechanism did not receive satisfying answers. Verworn (quoted by Woodger 1967, p. 238), for example, simply replied to them that "it is self-evident however, that only such laws as govern the material world will be found governing material vital phenomena – the laws ... [of] mechanics." Others admitted that this ideal of mechanism was a "dream and faith" (see Woodger 1967, p. 259). It is interesting to note that when Verworn made this statement on the self-evidence of mechanics, physics had already outgrown classical mechanics. How can one expect then that complicated living systems should be completely understandable in terms of mechanics, if the latter is not even sufficient to explain all aspects of nonliving matter! For that reason it is "fairly safe to say that so far as present-day biology goes, mechanism (in the sense of classical mechanics) ... is of very little significance, and when biologists talk about mechanical explanation ... they do not have anything like this in mind" (Woodger 1967, p. 262).

So, what do they have in mind? One of the following three notions:

(1) Living systems can and/or should be viewed as physico-chemical systems.

(2) Living systems can and/or should be viewed as machines. This kind of mechanism is also known as machine theory of life.

(3) Living systems "work in an orderly way" (de Beer 1924, p. 112, quoted by Woodger 1967, p. 258). This kind of mechanism "makes no assertions about the processes studied, but merely asserts that they take place according to laws" (Woodger 1967, p. 258). Hence, it may be compatible with materialistic, machine theoretical, organismic, and even some holistic and vitalistic philosophies of life.

Thus, when biologists enquire about the mechanism of a process they may simply think of the order or pattern that explains the process (definition 3). However, they may in addition imply that this mechanism is physico-chemical (definition 1) or machine-theoretical (definition 2). Hence, discussion concerning mechanism and mechanistic explanation may be ambiguous unless the meaning of mechanism is stated.

In order to evaluate mechanism, it is necessary to deal with reductionism because the latter is closely interconnected with the former and, furthermore, mechanistic doctrines are often formulated with reference to reductionism.

## 9.4 Reductionism

The two principal definitions of mechanism (definitions 1 and 2) can be stated in terms of reductionism as follows:

(1) Living systems are reducible to physico-chemical systems.
(2) Living systems are reducible to a machinery.

These definitions obviously require an understanding of what is meant by reduction. Ayala (1974a) distinguished three kinds of reduction: ontological, epistemological, and methodological reduction. This distinction eliminates at least some of the confusion in this field. Furthermore, it shows that some of the controversies and mutual accusations between reductionists and anti-reductionists are due to a lack of differentiation between different kinds of reductionism. "It is not untypical, for example, to see a reductionist concerned primarily with the ontological question accusing a self-proclaimed anti-reductionist of being a vitalist, while the latter may be an epistemological anti-reductionist but also in fact an ontological reductionist" (Dobzhansky et al. 1977, p. 488). Further complications may arise as a result of different definitions. For example, 'epistemological reduction' may have different meanings and other terms may be used (R.M. Burian, personal communication). The following discussion will, however, be based on the three kinds of reduction as defined by Ayala (1974). I shall try to explain the three kinds of reduction with reference to the first definition of mechanism according to which living systems are physico-chemical systems.

218

*Ontological reductionism.* "Ontological reductionism claims that organisms are exhaustively composed of nonliving parts (which are the atoms also found in inanimate matter). No substance or other residue remains after all atoms making up an organism are taken into account. Ontological reductionism also implies that the laws of physics and chemistry fully apply to all biological processes at the level of atoms and molecules" (Dobzhansky et al. 1977, p. 488). At first glance this postulate may appear to be rather straightforward and unproblematic, especially to the biologist who is used to thinking in physico-chemical terms. Upon further analysis it leads, however, to quite difficult questions such as the following: if living systems are solely composed of atoms and molecules, are they nothing but aggregations of atoms and molecules? In other words: is a living system nothing but the sum of its parts, specifically its atoms and molecules? If these questions are answered in the affirmative, it follows that the properties of living systems are those of their constituent atoms or molecules; or, more carefully expressed, the properties of living systems can be fully described in terms of the properties of their constituent atoms and molecules. Hence, a complete knowledge of the physics and chemistry of life would entail a full understanding of life.

Although some biologists may still subscribe to this naive form of ontological reductionism, many biologists and philosophers, who have reflected on this problem, agree that this naive reductionism is based on a fallacy which has been called the "nothing but" fallacy (see, e.g., Dobzhansky et al. 1977, p. 488). From the statement that organisms are solely composed of atoms or molecules it does not follow that therefore they are nothing but an assemblage of atoms or molecules. "A steam engine may consist only of iron and other materials (i.e., atoms or molecules), but it is [also] something other than iron and the other components. Similarly, an electronic computer is not only a pile of semiconductors, wires, plastic, and other materials" (Dobzhansky et al. 1977, p. 489). Stated in terms of properties this means that the whole system has properties that are not exhibited by its constitutet parts. These properties are called emergent properties; they are the result of emergence. Thus, as a more complex system arises from the interaction of more elementary units, emergence of novel properties occurs. Simon (1971, p. 155) mentioned the following examples of emergence: "the sound of two (or more) hands clapping, the texture of a woven fabric, the functional properties of a wire sieve, a geometrical pattern formed out of pebbles." As these examples show, emergence is not restricted to living systems but occurs at many levels of organization including inanimate systems. An often quoted example at the level of inorganic chemistry is sodium chloride (NaCl), common salt (see, e.g., Dobzhansky et al. 1977, p. 489). It has properties that are not exhibited by its constituent parts, which are sodium and chlorine. In fact the properties of sodium chloride are quite different from those of sodium and chlorine. Thus, already at the level of chemistry we find the phenomenon of emergence (Platt 1961). As the systems become increasingly complex, the results of emergence become more and more striking. Mind and consciousness may also be seen as the result of emergence (see below, Sect. 9.8).

Emergent properties result from the particular way in which the elementary parts are organized and integrated to form a system. Hence, organization and integration, which in turn reflect interaction, interdependence and interpretation, are key concepts for an understanding of life (see, e.g., Levins and Lewontin 1985). From this vantage point, I want to present some critical remarks concerning ontological reductionism:

(1) As pointed out already, ontological reductionism in its naive (nothing-but) form appears to be untenable because it disregards emergence.

(2) Ontological reductionism that allows for emergence is of no great use to further an understanding of living systems. Listing the component parts including the atoms and molecules of a living system does not explain its functioning. The crux and the challenge in the elucidation of living system is the interaction and integration of the parts, i.e., the problem of organization.

(3) It might be pointed out that, although a knowledge of the elementary units is not sufficient for an understanding of a living system, it is a prerequisite because how else could one study the interaction and integration of elementary units unless one knew those units. Thus atomism in a narrow and in a wide sense is seen as the absolute basis of any biology, regardless of whether naive ontological reductionism or emergentism is adopted. This tenet, although it may appear plausible, is, however, highly questionable (see, e.g., Levins and Lewontin 1980, 1985). Before criticizing it, the two kinds of atomism have to be briefly defined. Atomism in the narrow sense (which is the common sense) refers to the doctrine that the building blocks of the universe are atoms (in the sense of modern physics). Atomism in the wide sense implies that reality is composed of units. These units may be atoms (in the sense of modern physics), or higher (or lower) level units. With regard to living systems such units may be genes (genetic atomism), cells, organisms, species, or even nonmaterial units (for the latter see, e.g., Sheldrake 1981).

Returning now to atomism s. str., in connection with ontological reductionism the crucial question is: are atoms the elementary units for the emergence of the properties of life? The answer of modern physics is no. Atoms themselves are composed of elementary particles; hence, their properties are also the result of emergence. What are elementary particles? Are they particles or waves of energy? According to quantum mechanics they are neither one nor the other. So where are the most elementary units of reality? It seems there are none. According to Bohm (1980), the manifest order, which we observe in our ordinary state of consciousness, is based on an implicate order which is characterized by undivided wholeness. This means that reality is not basically atomistic. Davenport (1979), a developmental biologist, made corresponding remarks. He wrote (p. 2): "it was formerly expected that experience could be reduced to fundamental and irreducible components. However, the analysis of lower levels of structural reality has refuted this expectation." The same can be said of higher levels of organization. Cells, for example, are not absolutely real units of life. They are at best one aspect of it (see Sect. 2.1). We may conclude then that atomism s. str. and s. lat. is not tenable in an ontological sense. Consequently, ontological reductionism is not valid because we cannot reduce higher levels of organization to the units of lower levels when these units do not really exist.

As a result of modern developments in physics and to some extent also in biology, reality is not so much viewed any more as consisting of things such as particles, atoms, and higher level units (see, e.g., Bateson 1979; Bohm 1980; Levins and Lewontin 1980, 1985; Capra 1982; Davies 1983). Instead it is perceived in terms of "relational and organizational properties" (Hull 1974, p. 129 emphasized this with regard to biology). Thus, things have dissolved as ultimately real building blocks of the universe. We are left with no-thing.

Nothingness is often looked upon as something negative, especially by many people in Western culture. It is conceived as absence of the real and therefore as threatening. Yet there cannot be absence of reality. There can, however, be absence of things, in which case reality is not constituted of units, but is a continuous whole. No-thingness in this sense is the same as oneness (wholeness and unity).

The experience of no-thingness is central in Zen Buddhism. Under different names it occurs in many other ways of life and religions. Great mystics of all times in the East and the West have tried to communicate this experience or have preferred to remain silent knowing that words cannot convey no-thingness because they are always about some-thing.

I think it is highly significant that in our century science and ways of direct intuitive experience converge in no-thingness (see, e.g., Capra 1975, 1982; Zukav 1979; Bohm 1980; Hayward 1984). It would be naive to assume that the no-thingness of the Zen Master and the modern scientist are exactly the same. No-thingness in science is expressed in terms of something else. This raises the problem to what extent and in what form science can go beyond thinking in terms of things. In other words: to what extent can science become dynamic? (see Sect. 8.2). Thinking in terms of things, even when they interact in processes, is at least partially static inasmuch as the things as things do not and cannot change. Change in this view is only a reshuffling of things which may, however, lead to different emergent properties.

Although nothingness of the scientist and the Zen Master are not exactly the same[1], I think that they overlap and thus refer to the underlying unity of reality. Science cannot come as close to its as the Zen Master who transcends all dichotomies of thought and all fragmentation arising from it; yet science has allowed us to glimpse what most scientists of the nineteenth century could not even imagine.

*Epistemological reductionism* (also called theory-reduction) has been discussed extensively by philosophers as well as scientists (see, e.g., Nagel 1961; Hempel 1966; Woodger 1967; Ruse 1973; Hull 1974; Ayala and Dobzhansky 1974; Mohr 1977; Wuketits 1978). In contrast to ontological reductionism it refers not to reality itself, but to laws and theories about reality. Hence, "in biology the central question of epistemological (theoretical, explanatory) reductionism is whether the laws and theories of biology can be shown to be special cases of the laws and theories of the physical sciences" (Dobzhansky et al. 1977, p. 491).

Nagel (1961), Hempel (1966), and others have clearly stated what epistemological reduction means and requires: the conditions of derivability and connectability must be fulfilled. *Derivability* means that the theories and laws of one branch of science (let us say biology) must be derived as logical consequences from the theories and laws of the science to which it is reduced (let us say physics). The science that is reduced is called the secondary science, whereas that one to which it is reduced is the primary science. In our case the question is whether biology is a secondary science that can be derived from the primary science of physics. *Connectability* refers to the concepts employed and specifies that all the concepts of the secondary science (in our case biological concepts) must be defined in terms of concepts of the primary science (let us say physics) without loss of meaning. In other words: biological concepts must be translated into physical ones. Thus, epistemological reduction is conceptual and nomic (i.e., referring to laws) reduction.

Having stated what epistemological reduction means, one can ask now whether biology can be reduced to physics and/or chemistry. At the present

---

[1] Note that my verbal representation cannot be adequate because I cannot but refer to things such as "some-thing" or the "same thing." In discussing no-thingness, I make it a thing which is contrary to what I am trying to communicate. Obviously, we are here at the limits of verbal communication. Hence, the importance of direct experience.

the answer is a definite no. I think all informed biologists and philosophers agree on that. However, a number of biologists and philosophers believe that eventually biology will be completely reducible to physics and chemistry and thus become a branch of the physical sciences. Other biologists are less optimistic in this respect and some even claim that complete reduction of biology is in principle impossible (see, e.g., Polanyi 1968). To the latter that include organismic biologists, holists, and vitalists, the autonomy of biological science is unquestionably established.

Instead of speculating about the future, I think it is more profitable to look at the present situation. Biologists use a large number of biological concepts such as chromosome, cell, organ, organism, species, consciousness, etc. All of these concepts and many others will have to be defined in terms of physical and chemical concepts in order to fulfill the condition of connectability. Furthermore, laws and theories, such as the Mendelian laws or the synthetic theory of evolution, will have to be logically derived from laws and theories of physics and chemistry. The difficulties that confront us in this respect are immense (see, e.g., Ruse 1973; Hull 1974; Mohr 1977).

One biological discipline in which may reductionists have been hoping to be successful is genetics. The aim has been to reduce Mendelian (classical) genetics to molecular genetics. Classical genetics would then become a special case of molecular genetics. Instead of two theories or paradigms one would have only one, namely molecular genetics. Thus, we would be one step closer to the ultimate aim of science which is to establish the most general and comprehensive theory.

So far nobody has achieved this goal. Different reasons can be given for the failure (see, e.g., Ruse 1973; Hull 1974; Mohr 1977). One reason is the problem of translation (connectability) of biological concepts such as that of the "cell"; hence, Lindenmayer and Simon (1979) discuss the possibility of reducing Mendelian genetics to a theory that contains both molecular and biological concepts. Another reason is a lack of exactness, especially of molecular genetics. Mendelian genetics has been axiomatized by Woodger (1937, 1959). However, "a similarly strict axiomatization has not been performed so far in the case of molecular genetics" (Mohr 1977; see also Lindenmayer and Simon 1979). Hence, even the logical prerequisites for a derivation of Mendelian genetics are lacking at present.

Although the attempted reduction of Mendelian genetics to molecular genetics has not been successful, "there can be little doubt that much has been learned from what has been accomplished up to the present" (Dobzhansky et al. 1977, p. 495). At the biophilosophical level we have learned, for example, that what at first looked like a successful reduction was in fact a replacement of one theory or paradigm by another. In a theory replacement the original theory is able to provide kinds of explanations and predictions that the new one cannot give and therefore the original theory remains useful and valuable. Hence, there will be a place for Mendelian genetics as long as it has not been completely reduced to molecular genetics. And even if such a reduction should be possible in the future, Mendelian genetics may remain useful and valuable for purely practical reasons just as classical mechanics is still a useful and important field.

Some philosophers, historians, and scientists doubt whether any reduction has ever occurred in the history of science (see, e.g., Kuhn 1970; Mohr 1977, p. 97). Popper (1974, p. 260) pointed out that in the much quoted model cases such as the reduction of thermodynamics to statistical mechanics almost always an unresolved residue is left so that the reduction is not really complete but also a replacement. With regard to progress in science this means that the new theories do not fully contain older ones as special cases. Hence, the older theories retain a certain value of their own because they refer to a part of the domain that the new theory cannot reach. This may suggest that unification of science (e.g., biology, chemistry, and physics) is not possible through reduction, but must occur by other means (see, e.g., Maull 1977), if it is possible at all.

Facing the fact that reduction of biology to physics is unattainable at present and for many reasons probably also in the future, one may question whether it is desirable to concentrate our efforts on reduction. It may be easier and more successful to build general and comprehensive theories on the basis of General System Theory and other general approaches that are not reductionist (see, e.g., Koestler and Smythies 1969; Pattee 1973; Weiss 1973; Bertalanffy 1968, 1975; Bunge 1979b; Wuketits 1978, 1983). Such theories would comprise the physical and biological (and maybe even the psychological and social) sciences. As Hempel (1966) has pointed out long ago, they "might well be couched in novel kinds of terms ... that afford explanations both for phenomena now called biological and for others now called physical or chemical. To the vocabulary of such a comprehensive unifying theory, the division into physico-chemical terms and biological terms might no longer be significantly applicable, and the notion of eventually reducing biology to physics and chemistry would lose its meaning" (Hempel 1966, p. 106).

We have to keep in mind that a dichotomy between the physical and the biological sciences is a tacit assumption of the reductionist ideal in biology. Only if this assumption is accepted can we entertain the question of whether biology should be reduced to physics and chemistry. If we look upon natural science in a more unified way, the question of reduction need not even arise.

Questions are not given awaiting our answer. We create questions by a certain philosophical and/or scientific outlook and attitude. Instead of debating just the questions and the possible answers, it often might be more profitable to look at the assumptions that lead to the questions. The endless and tiring controversies over reductionism may very well become obsolete through more synthetic and comprehensive approaches that go beyond the traditional boundaries of disciplines.

If synthetic approaches should fail or – as I suspect will be the case – should be feasible only to a very limited extent, another fruitful alternative is perspectivism (Bertalanffy 1975) according to which different paradigms, theories, descriptions, etc. are seen as complementary perspectives of reality. Accordingly, reductionist and holistic accounts (i.e., descriptions at different organizational levels) may be complementary to each other (see, e.g., Varela 1979, p. 104; Hofstadter 1979; Davies 1983; p. 63; Lewontin et al. 1984, p. 280). For biological research this means "that different approaches are necessary to fully describe a particular living system, e.g., biochemical analysis of the elements, physiological blackbox experiments (input-output experiments), in vitro experiments with isolated organelles, microscopic analysis of the living system, light microscopic and/or electron-microscopic observations of the fixed and appropriately 'stained' system. Information from all these levels is required and indispensable in order to obtain a balanced description of the system" (Mohr 1977, p. 106). Even research at a given level requires different perspectives. Levins (1968) illustrated this with regard to model building in population biology. Since it is difficult to develop models that maximize generality, realism, and precision, we have to sacrifice any one or two of these ideals to the other(s). For example, we sacrifice generality to realism and precision, or we sacrifice realism to generality and precision. The conclusion of all this is the recognition of a pluralistic philosophy, i.e., the need for a plurality of approaches.

It is obvious that the discussion of reductionism is not only of theoretical and philosophical interest, but has also very practical and economic aspects as far as the distribution of research grants is concerned. Should those researchers whose aim is reduction receive higher grants than nonreductionists? The reader may form his own opinion in the light of what has been said by various informed philosophers and biologists.

*Methodological reductionism*, inasmuch as it concerns biology in terms of our first principal definition of mechanism, claims that living systems should be studied at the level of molecules and atoms. "For example, genetics should seek to understand heredity ultimately in terms of the behavior and structure of DNA, RNA, enzymes, and other macromolecules rather than in terms of whole organisms, the level at which the Mendelian laws of inheritance are formulated" (Dobzhansky et al. 1977, p. 490).

The counterpart to methodological reductionism has been called methodological compositionism (Simpson 1964) "which claims that to understand organisms we must first explain their organization ... Accordingly, organisms and groups of organisms should be studied as wholes, as well as in their component parts" (Dobzhansky et al. 1977, p. 490). The history of biology shows that reductionist as well as compositionist approaches have been of interest. Both may have heuristic value.

## 9.5 Machine Theory of Life

There are different forms of machine theory of life that may or may not be reductionist:

(1) One form of machine theory equates an organism with a machine. Accordingly, an organism is nothing but a machine (ontological reduction). The question that arises immediately is what sort of machine? A simple mechanical device such as a steam engine, a cybernetic automaton, or a highly sophisticated computer? Probably most, if not all, machine theorists have the two latter kinds of machine in mind. Although such a sophisticated machine shares many properties with living systems, it also differs from the latter in many respects. Therefore, a total equation of a living system with a machine is not possible. Any biologist who is not intuitively convinced that (s)he is not just a machine is advised to study the literature on this topic which provides different reasons why such a view is untenable (see, e.g., Bronowski 1966; Bertalanffy 1967; Boden 1979). I shall mention here only the argument advanced by Bronowski (1966). He explained with regard to the human mind that it is characterized by self-reference. Self-reference means that it refers not only to an object but at the same time to itself. Logical paradoxes such as the statement: "all men and women are liars" are a result of self-reference. In literature, it is of even greater importance than in logic and science. Bronowski pointed out that machines including the most sophisticated computers, lack self-reference. And he concludes: "All that we can say, and all that I can assert, is that we cannot now conceive any kind of law or machine which could formalize the total modes of human knowledge" (Bronowski 1966, p. 14).

(2) Another form of machine theory of life, which more appropriately is called the machine analogy (Woodger 1967, p. 259) or the computational metaphor (Boden 1979, 1981), is much more modest. It does not imply that a living system is nothing but a machine, but focuses simply on certain resemblances (analogies) between machines and organisms. One can hardly deny that modern machines such as robots or computers exhibit properties that characterize living systems. For example, they may show self-regulation, goal-directedness, memory, and other features of organisms. As a result, computers supplied with the appropriate programs can serve as models for living systems and thus simulate certain aspects of life.

Computer modeling and simulation have become increasingly important in modern biology and have led to fascinating insights. As I pointed out already, one classic example is the simulation of homeostasis. As more refined computers and computer languages are becoming available more life processes can be simulated. For example, Lindenmayer (1978, 1982), using the computer languages that he has developed during the last decades, has succeeded in simulating the development of intricate patterns of algae, fungi, and flowering plants, such as the form of branching, leaf arrangement, sequence of flowering, etc. Psychologists have been using the computational metaphor in cognitive psychology to simulate modes of perception, thinking, volition, etc. Some authors have called it the "computer revolution" (see, e.g., Sloman 1978). According to Boden (1979, p. 114) programs exist already that can do things such as the following: "perceiving in a holistic as opposed to an atomistic manner, using language creatively, ... distinguishing between different species of emotional reaction according to the psychological context of the subject." Boden (1979, p. 128) thinks that even "problems concerning human purpose, self, consciousness, freedom, and moral choice can be clarified by this metaphor" (see, however, also Kurzweil 1985).

The computational metaphor need not be reductionist in an epistemological sense and hence not mechanistic in that sense. Simulations may be written in terms of psychological or biological concepts. This is one reason why the computational metaphor is more comprehensive and more fruitful than epistemological reductionism. It is also more flexible: as certain biological or psychological concepts can be reduced to lower levels of complexity, low-level concepts can be used. For example, Lindenmayer (1978) in his simulations of leaf arrangements (phyllotaxis) employs the biological concept of "inhibitor." If the chemical nature of it were elucidated, the biological term could be reduced to or at least replaced by a chemical one.

In spite of the successes of the computational metaphor it is easy to point out limitations. These are admitted by those who utilize the metaphor critically (see, e.g., Boden 1979, p. 129). Some of them will probably be overcome by advances in computer software and hardware. Others may be inherent in the metaphor as well as the scientific approach (see, e.g., Heaton 1979). Varela (1979) pointed out that the computer approach is always one-sided and incomplete. If it is used alone, it leads to imbalance. Hence, complementary approaches are needed.

## 9.6 Organicism

Mechanism in the sense of the computational metaphor may come rather close to (or even coincide with) some forms of the biophilosophical outlook and approach that has been called organicism or organismic biology (see, e.g., Plamondon 1975). Authors who have developed this view of life include Bertalanffy (1952, 1967, 1968, 1975), Weiss (1967, 1968, 1973), and others (see, e.g., Blandino 1969; Haraway 1976). Needless to say, organicism, as most other general doctrines on life, has many versions and is therefore not easy to characterize. Beckner (1959, p. 5), in a detailed analysis of organicism, distinguished the following four principal doctrines of the organicist view of life:

1. organizing relations
2. directiveness
3. historicity
4. autonomy of biological theory.

1–3 are doctrines about living systems, whereas 4 refers to the theory of living systems.

To 1: The first doctrine refers to central issues of biology and biophilosophy, namely to organization, integration, and emergence. It excludes on the one hand mechanism in the sense of epistemological and methodological reductionism, and on the other hand vitalism that postulates a substantial vital principle in the form of an additional substance or force. It is in accordance with certain forms of holism and, although it does not deny that living systems can be looked upon as consisting of atoms and molecules, it need not be atomistic in an ontological sense. Hence, different interpretations of the meaning of organizing relations are possible. According to all of them the whole is more than the sum of its parts because the relations between the parts lead to the emergence of new properties. These new properties are not those of an additional entity but are those of the organization of the whole not found in the isolated parts. Only in this sense is the whole more than the sum of its parts.

To 2: Directiveness is understood by most organicists in terms of goal-directedness behavior (teleonomy) and functional behavior (see Chap. 7). Since teleonomic and functional behavior do not play an important role in the physical sciences, directiveness is seen as a phenomenon typical of living systems.

To 3: The historical aspect, which leads to the problem of uniqueness, is also characteristic of living systems. Beckner (1959) distinguished three postulates of historicity:

a) All organic systems have histories.
b) The past of an organic system determines, or helps to determine, its present structure and behavior.
c) Much of the historical development is irreversible.

228

Several of the issues that may arise from these statements have been discussed or alluded to in Sect. 8.3.

To 4: The doctrine of the autonomy of biological theory implies anti-reductionism in the epistemological and methodological sense. It may be compatible with at least one form of ontological reductionism according to which the laws of physics and chemistry apply to living systems and the latter consist solely of atoms and molecules. However, if that is granted, it is emphasized that additional higher-level laws are not reducible to physics and chemistry because they refer to biological organization that does not exist at the physico-chemical level (see, e.g., Polanyi 1968; Ayala 1968; Wuketits 1978, 1983). Inasmuch as the laws of physics and chemistry are valid for living systems, the term 'autonomy of biology theory' may be slightly misleading because biological theory is autonomous only in terms of biological laws, whereas with regard to physico-chemical laws it is not autonomous. This is the prevalent attitude of organicists.

As Bertalanffy (1952, 1967, 1968, 1975) pointed out, organismic biology (or organicism) transcends both mechanism (in terms of epistemological and methodological reductionism) and vitalism (in terms of a substantial vital principle). In that sense it is a synthesis at a higher level and a considerable achievement of theoretical biology and biophilosophy. It should be clear, however, from the discussion in the preceding chapters that all doctrines of organicism pose problems that await solutions. Nonetheless, a definition or characterization of life can now be attempted in terms of organicism which includes the physico-chemical basis.

### 9.7 What is Life?

So, what is life? It could be said that it is the characteristic of living systems. The question then is: what is a living system? The following answer might be given:

Living system = an open system that is self-replicating,
self-regulating, exhibits individuality, and
feeds on energy from the environment.

The openness of living systems is a fundamental feature and very typical, although not restricted to living systems. A flame, for example, is also an open system. For that reason, Bertalanffy (1967) referred to life as the 'living flame.' Openness is thus a necessary condition for the definition of living systems, but not sufficient.

Openness of living systems exists with regard to material as well as energy exchange with the environment. The implication of this openness is that no absolute boundary can be assigned to a living system. Any boundary that we draw is relative. This means that living systems do not exist as totally separate entities: they are integrated in more inclusive wholes. Thus, organisms are integrated into ecosystems which in turn are integrated into the whole planet and universe (see, e.g., Lovelock 1979).

I think that the discovery of the openness of living systems has been one of the most important contributions of modern biology at least as far as existential and social consequences are concerned. If we could become totally aware of this natural openness, our human existence as well as society would be transformed. Too often we behave as if we were isolated boxes and thus we act against nature. As a result of this unnatural behavior conflicts and diseases of all sorts arise and we thus find ourselves in the predicament of modern society (see, e.g., Krishnamushi 1970; Krishnamurti and Bohm 1985).

Weiss (1973) emphasized the relevancy of openness for education. As education becomes less rigid and dogmatic, young people grow up in a more open and tolerant way. The consequences for society as a whole are beneficial and liberating. One consequence of utmost importance would be the realization that seemingly opposing views and groups sharing them are not always as fundamentally different as it is thought. The difference is exaggerated due to the abstraction of a few differing features and the formation of categories on the basis of those selected features. Very often it is overlooked that a continuum exists between the so-called typical representatives of the categories. If one takes into consideration this whole continuum, the opposing camps as distinct entities are seen to be nonexistent: the supposed dualism turns out to be unreal. Thus antagonisms, which were built on supposed ultimate difference, lose their ground. Opposing views and groups sharing them can be seen to be open toward each other through a continuum that forms a whole instead of mutually exclusive opposites.

It has often been said that we should build bridges between views and people that are in opposition. In this way we could increase peaceful co-existence and harmony. I think bridges are already there but we are unaware of them. What is needed is increased awareness and insight.

Some awareness has been generated already by science including biology as well as the counterculture movement. Since the latter often shows anti-scientific tendencies, it is frequently seen in opposition to science. Here again, a dualism has been erected that does not exist. Those branches of modern biology that deal mainly with integration and unity at a global level come close to holistically minded representatives of the counterculture movement whose major concern is also the realization of the whole in contrast to the blind worshipping of fragments and artefacts. There may be considerable difference between biological holism and the holism of the counterculture both of which are again very heterogeneous in themselves. But in spite of

230

differences, the two overlap and in the region of overlap refer to that aspect of reality that unites opposites.

Since the counterculture movement is strongly influenced by oriental philosophies and religious, the convergence between our Western culture and its counterculture reflects a convergence of Western science and Eastern wisdom (see, e.g., Capra 1975, 1982; Snyder 1978; Bohm 1980; Wilber 1982). Again, East and West are not absolutely in opposition to each other. They are open toward each other and also form a continuum. Eastern people also use their intellect in fragmenting and destructive ways as people in the West. On the other hand, Western people have wisdom too. Profound wisdom is probably rare in the East as well as in the West.

In spite of resemblances between West and East, or Western culture and its counterculture, differences remain, especially if one looks at the extreme ends of the continuum. Not seldom external differences in the dress and behavior of people reinforce the belief in absolute differences. For example, biologists in their white laboratory coats surrounded by sterile equipment form a striking contrast to critics of our Western society who live a simple life in a mountain retreat chanting Buddhist sutras and doing organic farming. There are, however, counterculture people who wear suits, white shirts and ties and there are scientists who dress like hippies. Hence, even with respect to external appearance we find a continuum between representatives of Western culture and counterculture, or West and East.

Let us now return to the definition of living systems. Since openness is only a necessary condition for the definition of living systems, other features are required that are necessary and sufficient. Self-replication may be considered both necessary and sufficient, although there may be systems that do no more reproduce and yet are still alive at a certain period of time. To avoid this problem one could refer directly to the chemical basis of self-replication, namely DNA and RNA. In this way it might be easier to characterize life on our planet, but one might exclude extraterrestrial forms of life (if they exist) that might have a different basic chemistry.

Self-replication and self-regulation are very characteristic features of living systems. They entail organization (see, e.g., Varela 1979). The latter becomes increasingly important in higher forms of life due to the complexity of networks within the system and between the system and its environment. Individuality implies that no two organisms are exactly alike.

Another basic feature of living systems is that they extract energy from the environment. They "create their ordered microcosm at the cost of their surroundings ... and live, so to speak, by extracting order or, as Schrödinger (1944) phrased it, they feed on negative entropy or negentropy" (Mercer 1981, p. 46). More specifically, the constant generation of entropy in living systems "is offset by the negative entropy brought in by the energy flow"

[Mercer (1981), p. 140; see, however, also Tonnelat (1978, 1979), Prigogine and Stengers (1979), and Trintscher (1973) who thinks that the concept of entropy does not apply to living systems].

The characteristic features of living systems mentioned thus far may be conceptualized differently, and additional features may be added (see, e.g., Varela 1979; Bretschneider et al. 1982; Wuketits 1983). Elsasser (1975) contrasts biological systems with physical systems in the following way:

Physical system: simple, unambiguous, quantitative, universal
Biological system: complex, varied, contingent, individual.

In a more recent summary of the principles of a biological theory, Elsasser (1981) states three basic postulates: (1) finiteness and, related to it, individuality, (2) creative selection which "implies that out of an immense number of states available the organism uses only an immensely small subset" (p. 145), and (3) information stability which refers to the relative stability of biological organization over enormous periods of time (see also Thom 1975).

Any general definition that is supposed to apply to all living systems by necessity must be most abstract and hence leave out most of the features of complex organisms including the human species. One of those features that in this way is excluded is mind and consciousness. Even if we would ascribe some proto-form of consciousness to all living system (see, e.g., Rensch 1970), we still would have to leave out the highly evolved human mind.

Critics of a general definition of living systems may point out that in a very strict sense a monotypic definition as the above can never apply to all forms of life at all periods of time. They may emphasize that any monotypic definition is either too wide so that it includes some nonliving systems or too narrow so that it excludes some living systems at least at some stages of their life. A feature that is too wide is that of openness because, as I mentioned already, there are certain nonliving systems such as flames that are open. Features that may be too narrow are those of self-replication and self-regulation because certain living systems may lack them at least at certain periods of time. Defining living systems polytypically instead of monotypically may improve the situation, but I doubt whether it would solve the problem completely. Hassenstein (1971) pointed out that life must be seen as an injunction (i.e., a concept that cannot be defined by a set of properties).

Life is a process and the features that have been used to define living systems refer to processes. We know, however, that there are living systems that may have resting stages in which life-processes are slowed down or may come practically to a standstill. This is the case for certain spores and cysts. Are these spores and cysts still alive? Perhaps one could argue that there is only a quantitative difference between these relatively inactive stages and the more typically living systems. Becquerel (quoted by Jeuken 1975) desiccated "various species of Rotatoria, Tardigrada, algae spores of bacteria and yeast, fragments of lichens and mosses, and seeds of various plants. Next he cooled them down to a temperature of approximately 0.01 K (this is 0.01° above absolute zero at −273.15 °C) and held them in this state for 2 h. Then he brought them gradually back to normal conditions, and all the organisms revived" (Mohr 1977, p. 105). Saying that they revived implies that they were dead while being cooled down. It is generally thought that very close to a temperature of absolute zero life-processes cease completely. A definition of life in terms of processes would mean death and revival in that particular instance. We have to keep in mind, however, that this statement is a scientific hypothesis.

Since there is a continuum between living systems in typically dynamic states and those in which life-processes cease nearly or completely, the notion of an injunction or fuzzy set once again provides a solution. In terms of these notions, the definition of a living system is fuzzy, i.e., systems at any particular time can have from 100% to 0% membership in the fuzzy set of living systems, which may be characterized by properties such as those utilized for the above monotypic definition.

If we assume now that, using fuzzy-set theory or the notion of injunction, we can arrive at a satisfactory definition of "living system," which can be interpreted as a well-corroborated theory of life, does that mean that we have answered the question: what is life? I do not think so. We would have given a scientific answer. This answer implies only that we have selected (i.e., abstracted) a few properties that characterize the whole spectrum of living systems. All the other properties and aspects of life, including those that cannot be conceptualized but only experienced intuitively and introspectively, are left out (see, e.g., Jeuken 1975).

Essentialists would argue that, although life is defined only by selected properties, these are the essential ones which, according to them, characterize the true nature of life. The other properties are nonessential or accidental. Such a postulate raises more problems than it solves. For example, how do we decide which properties are essential? If those are essential that are shared by all members of the class, we would arrive at absurd conclusions. This would mean, for example, that profound scientific or intuitive insight is nonessential because it is so rare, whereas ignorance, which is much more widespread, would come much closer to be an essential feature, at least of humankind. (S)he who finds this analogy inappropriate may consider the following. The so-called essential properties are dependent on the so-called nonessential ones. The former cannot exist without the latter. Hence the so-called nonessential features of a living organism are "essential" to sustain life.

I think that essentialists have a biased outlook. They select the so-called essential features, disregard or neglect the remainder and thus lose sight of the whole. This amounts to a fragmentation of nature which may be destructive and dangerous for the individual as well as society.

One aspect that also tends to be neglected or excluded by an essentialistic definition of life is feeling. Feeling life may reveal just as much of life as thinking about it and analyzing it scientifically. In fact, feeling may reveal more about life than thinking. Some would go further and claim that feeling (or intuition) alone reveals life as it is, and thus answers the question: 'what is life?' I think it may be questionable and dangerous to look upon feeling as the absolute judge of truth. Feeling like thought is usually conditioned by our upbringing, tradition, and culture. Furthermore, it may be influenced by thought so that what is supposed to be pure feeling is at least to some extent a reflection of thought (see, e.g., Heaton 1979, p. 183). Feeling, then, at least for the vast majority of people of all cultures, presents only one perspective of life, though a very important and basic one.

Life as it is may be beyond our experience except to those very few totally liberated and enlightened beings who, because of a total awareness, can experience and live life as it is in its suchness. Needless to say, such experience is beyond all dichotomies of discursive thought including the dualism of life and nonlife. Reality is experienced as oneness, or one might say, if one wants to continue using words where they are inadequate, reality is experienced as the "Manifold and the One" (Arber 1957).

Although science due to its limiting methodology cannot reach that insight, it is most impressive how close modern science has brought us to a glimpse of oneness and unity which, as I pointed out above, is not the same as that of mystical experience of oneness but touches upon it or overlaps with it to some extent inasmuch as it refers to the unity of the universe (e.g., Barash 1973). Modern ecology, for example, has described the integration of living systems with the so-called abiotic environment, emphasizing that both form a unity which has been termed ecosystem. In a sense, this notion of ecosystem transcends the naive belief that living systems are absolute entities that in reality exist separately from nonliving systems. Bateson (1972) has gone one step further in his "ecology of mind" by pointing out that because of the wholeness of reality mind is also integrated with nature in such a way that it is all-pervasive. In that sense Bateson (1972) said that Lake Erie is part of our mind. "If Lake Erie is driven insane, its insanity is incorporated in the larger system of *your* thought and experience" (Bateson 1972, p. 484).

If living systems are seen as integrated with nonliving nature into more inclusive systems such as ecosystems, the question of 'what is life?' loses some of its fundamental significance or is seen in a vaster perspective. Even if life could be defined in isolation from nonliving matter, it cannot be com-

234

pletely understood this way because it is integrated with so-called nonliving nature. To understand it more completely, one will have to look at it from the vaster perspective of more inclusive systems.

## 9.8 The Mind-Body Problem

A thorough discussion of the relation between mind and matter, mind and body, and soul and body is beyond the scope of this book. The reader is referred to the very extensive literature (e.g., Woodger 1967; Rensch 1970; Ayala and Dobzhansky 1974; Popper and Eccles 1977; Bunge 1980; Rieber 1980; Bohm 1980; Hofstadter and Dennett 1981; Wilber 1982). Yet I shall make at least a few remarks on the mind-body problem, which is not only an important issue in the philosophy of psychology, but also of interest to the biophilosopher. It is one of the age-old riddles of philosophy and science and can be stated in many forms, since it has many dimensions. For example, it can be presented in terms of the relation of two kinds of knowledge:

(1) public knowledge such as the facts and general statements of natural science which we share with others; for example, the fact that "this frog has four legs" or the generalization that "all brains consist chiefly of nerve-cells and processes" (Woodger 1967, p. 459).

(2) private knowledge which is obtained through introspection; for example, the fact that "I am now thinking about the next general election" (Woodger 1967, p. 459).

Expressed more concretely, the mind-body problem presents itself as follows. We can describe in biological and physico-chemical terms how light impinges upon the retina of our eyes, the biochemical and electrical processes occurring there and during the conduction through the optical nerve to the brain. And then at some point we experience a sensation that like thought is of a different kind from physico-chemical and biological phenomena. What is the connection and/or relation between these two kinds of phenomena? This is another formulation of the mind-body problem. Bunge (1980, pp. 3, 211) distinguished ten different views on the mind-body problem. I shall mention only four of them:

(1) *Interactionist view.* According to this view, mind, and body are different kinds of realities which interact with each other. The problem is, of course, how such an interaction is possible or even conceivable. No satisfactory answer has been provided as yet. For example, Popper and Eccles (1977, p. VII) who subscribe to this view "think it improbable that the problem will ever be solved, in the sense that we shall really understand this relation."

(2) *Psycho-physical parallelism.* According to this view physical (bio-logical) and mental events run side by side without mutual interaction. The advantage of this view over the first one is that it circumvents the problem of how something physical may become mental or vice versa. One question that arises, however, is how such a parallelism is possible without any inter-action. What makes the physical and mental events coordinated in a parallel fashion? And do all physical events have mental counterparts? If not, what determines whether a physical event has a mental parallel?

Since, according to psycho-physical parallelism, the chain of physical events is never broken by a transformation into something nonphysical (as the interactionists postulate it), physical reality is self-contained and can be understood independently of the mental parallel. From here it is only one step to consider mind as an epiphenomenon of matter or to deny its existence altogether and thus arrive at a materialistic world view. Alternatively, one could be led to the belief that it is only the mental or spiritual that really exists (e.g., various idealistic philosophies).

(3) *Identity postulate* (or double aspect views). According to this view, mind and matter are not different realities, but only different aspects of the same reality. Hence, the basic dualism of matter and mind is overcome and thus the problem of the interactionists appears as a pseudo-problem: if mat-ter and mind are aspects of the same reality, the same "common ground" (Bohm 1980, p. 209), they cannot interact because for an interaction two things are required. Psycho-physical parallelism as well as its degeneration into materialism or idealism are also superseded by this identity view.

(4) *Emergentist materialism.* This view, like the identity postulate, is monistic in the sense of psycho-physical monism, i.e., the physical and the mental are considered to be one. The mental is, however, seen as a function of the material: as the material reaches the systemic complexity of the human brain (or that of highly evolved animals), mind emerges (see, e.g., Bunge 1980). Thus, mind is a function of the brain, the latter being quali-tatively different from other material systems such as computers (Bunge 1980, p. 2).

This view of emergentist materialism goes far beyond naive materialism [e.g., the view that mind is an illusion or nothing but a swarm of particles (see Bunge 1980, p. 5)]. However, it takes the primacy of the material for granted. It equates reality with the material, i.e., "reality is composed ex-clusively of material or concrete things" (Bunge 1980, p. 224). In contrast, according to the identity postulate, reality is the Unnamable and the "mate-rial" is an abstraction from the unnamable reality.

One consequence of the emergentist view is that there are as many minds as there are brains. In contrast, according to the holographic paradigm (see Wilber 1982), mind can be singular, thus reflecting the basic organization of the universe including all existing brains (Pribram 1982). "In the implicate, holographic domain, the distinction between points becomes blurred; information becomes distributed as in the example of the surface of a pond. What is organism (with its component organs) is no longer sharply distinguished from what lies outside the boundaries of the skin. In the holographic domain, each organism represents in some manner the universe" (Pribram 1982). One advantage of this view is that it provides a conceptual framework that may provide an explanation for parapsychological phenomena: in the implicate order of the holographic domain neither space nor time exist; it is characterized by "undivided wholeness in flowing movement" (Bohm 1980, p. 11). Another merit of this holographic view is that it points to the underlying unity of all "things" that are not absolutely real and disconnected from each other and therefore do not exist as entities (see Bohm 1980). "Basically, everything is one" (McClintock, quoted by Keller 1983, p. 204).

## 9.9 Summary

*Introduction.* The meaning of the question 'what is life?' is not given in a neutral fashion, but reflects prior philosophical assumptions of the person who is asking the question. These assumptions may be conscious or more often partially or totally subconscious. If a dualism between life and inanimate matter is assumed, essentialism may be implied. In a nondualistic view of reality, life may be considered to be reducible to nonliving matter or living and nonliving matter may be seen as an integrated unitary system.

*Vitalism.* Vitalism may refer to the belief that the whole cosmos is animated (animism), or, more specifically, to views that life is infused by a vital principle which is absent in nonliving matter and which has been conceived of as a vital substance, fluid, or force. To adherents of the latter view, the answer to the question 'what is life?' is simple. Life is whatever is governed by the vital principle. Four kinds of criticism of vitalism are described. Although all of them are valid to a certain extent, they need not apply to all forms of vitalism. Not much is gained by condemning or ridiculing vitalism and all its adherents. Such condescendence is often based on ignorance. Vitalism, including its modern neovitalistic versions, is extremely heterogeneous and at least in some of its forms has the merit of drawing attention to aspects of life that have been neglected by other doctrines of life.

*Mechanism.* Besides an outdated definition of mechanism, according to which biology should be reduced to classical mechanics, three meanings of mechanism can be distinguished in modern biology. According to them, living systems can and/or should be viewed (1) as physico-chemical systems, (2) as machines, and (3) as orderly systems that exhibit lawfulness.

*Reductionism.* In terms of reductionism, definitions 1 and 2 of mechanism can be stated as follows: (1) Living systems can and/or should be reduced to physico-chemical ones, (2) Living systems can and/or should be reduced to machinery. Three forms of reduction have to be distinguished: ontological, epistemological, and methodological reduction. These three forms are discussed with regard to physico-chemical reduction. Ontological reductionism in its naive form states that a living system is nothing but an assemblage of atoms or molecules. In a more respectable form, ontological reductionism acknowledges that at higher levels of organization properties emerge that do not exist at the level of atoms and molecules. Emergence of these novel properties is the result of the organization and integration of the lower-level units. Whether units in the form of things exist in nature has become increasingly doubtful and thus atomism s. str. (referring to atoms) as well as atomism s. lato (referring to units at any level) are debatable. Nature and life are now not so much viewed as consisting of things but conceived in terms of relational and organizational properties. In this way modern science leads to no-thingness, which has been a central experience to mystics of all times in East and West. Although nothingness in science cannot be exactly the same as its experience by an enlightened person, a certain overlap and convergence of an idea and an experience appears to be a highly significant event of the twentieth century.

Epistemological reductionism requires: (1) the logical derivation of biological laws from laws of physics and/or chemistry, and (2) the translation of biological terms into physico-chemical ones. It is generally agreed that this is not feasible at present. Even the reduction of Mendelian genetics to molecular genetics has failed thus far. Whether any such reduction will be possible in the future is questionable. It is even questionable whether the postulate of reduction is desirable. Instead of trying to reduce biology to physics, one could aim at developing unifying theories that encompass the laws and concepts of biology, chemistry, and physics. Alternatively, different approaches may be seen as complementary to each other (perspectivism, pluralism).

Limitations of methodological reductionism are pointed out. This form of reductionism demands in contrast to compositionism that living systems are studied only at the physico-chemical level. Practical, economic, and political implications of all sorts of reductionism are mentioned with regard to the problem of how research grants should be distributed.

*Machine theory of life* in its naive form claims that living systems are nothing but machines. In a more modest form it is known as the machine

238

analogy or the computational metaphor whose aim is to use sophisticated computers as models for living systems. This approach of simulation need not be reductionist and is more flexible than mechanism in a physico-chemical sense.

*Organicism* has been characterized by the following four doctrines: (1) organizing relations, which imply that the whole is more than the sum of its parts (emergence), (2) directiveness (i.e., teleonomy), (3) historicity, and (4) autonomy of biological theory, i.e., epistemological anti-reductionism.

*What is life?* This question may be answered by stipulating that it is a characteristic of living systems which are then defined as open systems that are self-replicating, self-regulating, exhibit individuality, and feed on energy from the environment. Other features could be added to this list of defining features, but such additions might worsen the general problem of any monotypic definition of living systems in terms of a set of properties, namely the problem that such a definition is either too narrow or too wide. A polytypic definition or the use of fuzzy-set theory are mentioned as better alternatives. However, even the best definition does not completely answer the question 'what is life?' because it leaves out many aspects of life, particularly those that can only be experienced intuitively and introspectively. Essentialistic definitions of life are criticized as biased, fragmenting, destructive, and dangerous. A realization of the openness of living systems could lead to a profound transformation of the individual as well as society. It also leads to the insight that even if life could be defined in isolation from nonliving matter, it could not be understood separately because it is integrated with the so-called abiotic environment to form a more inclusive whole at which we have to focus our attention if we want to gain more profound insights and live in harmony with nature. Even those whose prime concern is just survival may have to look at the whole. What thus may have been intended to serve mainly utilitarian purposes may turn out to create a world of more awareness and peace.

*The mind-body problem.* Bunge (1980) distinguished ten different views on this problem. Four of them are briefly mentioned: (1) the interactionist view which holds that mind and matter are two different realities, the problem being how and where they interact, (2) psycho-physical parallelism which claims that physical (biological) and mental events run side by side without interaction, (3) identity of mind and matter so that the two are viewed as different aspects of the same reality; according to this view the interactionist problem appears as a pseudo-problem and psycho-physical parallelism is superseded as well as the two philosophies into which it has degenerated, namely materialism and idealism. (4) Emergentist materialism considers mind as a function of a high degree of complexity and systemic organization of the material. One consequence of this view is that there are as many minds as there are (highly evolved) brains. The oneness of the universal mind, which according to the holographic paradigm reflects the basic organization of the universe including all brains, cannot be grasped from the point of view of emergentist materialism.

# 10 World Hypotheses (World Views) and Truth

"Life itself has to be regarded as belonging in some sense to a totality ... even when it is not manifest, it is somehow 'implicit' in what we generally call a situation in which there is no life" (Bohm 1980, p. 194).

"To understand the whole of us and the world, we have to participate with the whole of us. Specifically, the bringing together of verbal and non-verbal forms of knowledge, rational and intuitive, is necessary" (Varela 1976).

"The world is a mystery ... it is no at all as you picture it ... well, it's also as you picture it, but that's not all there is to the world; there is much more to it" (Castaneda 1974, p. 165).

## 10.1 Introduction

As pointed out in the preceding chapter, living systems cannot be understood in isolation from their environment because they are open toward their environment and integrated with the latter. Therefore, understanding life requires and understanding of the organism and its environment which is the whole world. This means that we have to include in our discussion of life hypotheses and theories that refer to the whole world. Pepper (1942/70) called such hypotheses and theories *world hypotheses*, using the term 'hypothesis' in the widest sense that includes "theory." According to Pepper, a world hypothesis is a comprehensive world view that is corroborated by facts: "every consideration is relevant to a world hypothesis and no facts lie outside" (Pepper 1942/70, p. 1). Although facts are fundamental, a world hypothesis also reflects a certain way of thinking. Since there are many different ways of thinking, we have a great variety of world hypotheses. In an attempt to cope with this diversity, Pepper reduced it to seven principal world hypotheses (see below). Although this reduction as any other one is debatable, it appears to be useful for a brief discussion of world hypotheses, provided we keep in mind that many of the original and unique ideas of individual authors of world hypotheses have been lost in the process of reduction.

Inasmuch as world hypotheses represent a certain way of thinking, they may influence the formulation of more specific hypotheses and even the perception of facts. Conversely, the empirical component of facts and lower level hypotheses may have an impact on world hypotheses. Thus, there is a feedback between world hypotheses and lower level hypotheses including facts. The following discussion will show that even the scientific questions we ask may be influenced by the world hypotheses. For example, the question 'what is life?' may have a different meaning depending on which world hypothesis is presupposed. Because world hypotheses play such a fundamental role, an inquiry into them is of general importance, even for the biologist who deals only with very specific and supposedly concrete problems.

## 10.2 Pepper's Seven World Hypotheses

Pepper's seven world hypotheses are the following: (1) the generating-substance hypothesis, (2) animism, (3) mysticism, (4) formism, (5) mechanism, (6) contextualism, (7) organicism. Pepper rejected the first three as inadequate and retained the latter four as "the relatively adequate hypotheses." I shall first refer to Pepper's relatively adequate hypotheses.

### 10.2.1 Formism

The central category of formism is that of form in the sense of eternal form. An eternal form is characterized by its essence. Any particular thing or entity represents or partakes in an (eternal) form. Those things or entities that are essentially similar (or essentially the same) belong to the same form or essence; or it is said that they represent the same plan, design, or norm. The fact that things or entities belong to the same form is indicated by the same name. For example, whatever is called red has the essence of redness, i.e., belong to the form of red.

According to Pepper, formism includes "realism" and "platonic idealism," and it is associated with Plato, Aristotle, and realists in general. I think that even in modern biology it is much more widespread than is generally acknowledged (see, e.g., Chaps. 4 and 7). I have criticized formism already under the names of 'essentialism' and 'conceptual realism' (Chap. 4) and have pointed out that it is a static and fragmenting philosophy that is not adequate to the dynamics of life (see also Pepper 1942/70, p. 185). As the rigid framework of static eternal forms, essences, plans, or norms is imposed onto the fluidity of life and human society, distortion and destruction may become severe and painful. The history of Western science and culture up to the present time is full of atrocities of formism. It is time that we realize this more fully to avoid more conflict, destruction and war (cf. Sect. 9.7).

Much of the scientific, philosophical and even religious discussion on the "nature" of life is more or less determined or influenced by formism, for the "nature" of life is often equated with the essence or form of life. Hence, the question may be to conceptualize (if possible in terms of a definition) the essence of life. This question presupposes that life or anything else has an essence of its own and thus is distinct from nonlife. In other words: it presupposes the validity of formism. Once this is accepted either consciously or subconsciously, then the question on the nature of life (in terms of its essence) becomes meaningful and by necessity rigid and fragmenting doctrines can be proposed which, again by necessity, omit most of the features of concrete living systems as nonessential. Consequently, the meaning of life becomes extremely impoverished and living systems are seen in isolation from their context. With regard to human life, we end up with symptoms so typical of modern society: isolation, loneliness, fear, impoverishment, etc.

## 10.2.2 Mechanism

In mechanism, so it is said, only particulars exist. In this sense mechanism is nominalistic. Hence, there are no forms distinct from the reality of the particulars. According to Pepper (1942/70, p. 193) the following categories are basic to mechanism:

Primary categories
1) Field of locations
2) Primary qualities
3) Laws holding for configurations of primary qualities in the field (primary laws)

Secondary categories
4) Secondary qualities
5) A principle for connecting the secondary qualities with the first three primary or effective categories
6) Laws, if any, for regularities among secondary qualities (secondary laws)

What does that mean? Any particular must be assigned to a location in the spatiotemporal field. Its primary qualities are size, shape, motion, solidity, mass (or weight), and number. These basic concepts have a long tradition, from Democritus to Galileo to modern science and naturalistic philosophy (Pepper 1942/70, p. 204). They are related to each other by laws. It is important to note that in what Pepper (1942/70, p. 195) called discrete mechanism many of the structural features of nature are treated in a fragmenting way as it is the case in formism. "So space is distinct from time, the primary qualities are distinct from the field of locations, each primary quality is perhaps distinct from every other, certainly every atom (i.e., localized group of primary qualities) is distinct from every other atom, has independence of its own, and every natural law (such as the law of inertia, or the law of action and reaction) is distinct from every other law, and distinct, moreover, from the field of locations and from the atoms distributed over the field" (Pepper 1942/70, p. 196). It is obvious that it is an enormous danger for mechanism to fall into the trap of formism and it is probably no exaggeration to say that many mechanists have become more or less infected by formism (see Pepper 1942/70, p. 210, for an example). The only way to avoid this danger of becoming formist is to imbed the primary qualities and laws in the spatiotemporal field, i.e., anything is real only if it has a time and a place (Pepper 1942/70, p. 211). This leads to what Pepper (1942/70, p. 212) called consolidated mechanism. Here "in place of the discrete particle is the spatiotemporal path, and in place of the discrete laws of mechanics is a geometry, or, better, a geography. The purpose of this cosmic geometry is simply to describe to us the unique structure of the spatiotemporal whole" (Pepper 1942/70, p. 212; see also p. 214).

Secondary qualities are those with which the biologist is usually concerned, such as cells, organisms, modes of behavior, consciousness. One could go further in saying that probably all the characters of human perception are secondary (Pepper 1942/70, p. 215). The burning problem for mechanism is how to reduce the secondary qualities and laws to the primary ones. The difficulties, if not the impossibility, of such a task have been discussed above (Sect. 9.4). This limits the scope of mechanism enormously because it leaves many, if not most, of the emergent secondary qualities of living organisms unaccounted for. Still more fundamental philosophical inadequacies are discussed by Pepper (1942/70, pp. 220, 230, etc.).

The preceding discussion of mechanism is so general that it includes all of the different meanings of mechanism distinguished in Sect. 9.3 with the partial exception of the definition of mechanism as the lawful conception of life. The belief in lawfulness characterizes not only mechanism, but also formism and organicism. According to Pepper's scheme the mechanistic world hypothesis comprises "naturalism," "materialism" (if the secondary categories are ignored, or denied, or taken as secondary), and perhaps "realism." It is associated with Democritus, Galileo, Descartes, Hume, Reichenbach, and others (Pepper 1942/70, p. 141).

As far as the question of life is concerned, mechanistic answers entail the attempted reduction of life to the primary mechanistic categories. This aim of reduction, if it is practiced dogmatically, may lead to an impoverishment of biological theory. This does not mean that mechanistic approaches are useless. As long as their limitations are seen, mechanistic interpretations may increase our understanding of living organisms. However, the question facing us is to what extent, if any, it will be possible to extend the scope of mechanism. Another question is in what direction mechanism as a philosophy will be able to evolve so that it may be able to overcome some of its principal obstacles. For some time already, mechanism has been intergrading not only with formism but also with contextualism and organicism. Since the latter two world hypotheses provide far greater scope than traditional mechanism, an amalgamation of mechanism with these philosophies may provide solutions to some of the dilemmas. Especially the computational metaphor (machine theory) is promising in this respect (see Sect. 9.5). Pepper (1942/70,

p. 147) pointed out that mechanism and contextualism tend to combine. He also noted that the two theories are in many ways complementary. I think similar remarks could be made with regard to mechanism and organicism.

Mechanism and formism are analytical, whereas contextualism and organicism are synthetic. The former often stress the reduction to basic elements or factors, whereas the latter emphasize wholes with their emergent properties and high degree of integration.

### 10.2.3 Contextualism

In contextualism the emphasis is on events in their actuality. However, events are not considered in isolation but in their context. In contrast to formism (and to some extent also to mechanism), this philosophy is highly dynamic. Change and novelty are of fundamental importance, more so than in any other of Pepper's four relatively adequate world hypotheses. Absolute permanence does not exist according to this theory. As a consequence many features are not universal and order is hard to come by. Disorder is accepted, yet order is not ruled out (Pepper 1942/70, p. 234). Events are integrated with each other and their context through a mesh of strands. Because of this great emphasis on integration, which leads to the emergence of more inclusive wholes, contextualism has a tendency toward organicism. As a result, one might even interpret contextualism and organicism as two versions of one world hypothesis (Pepper 1942/70, p. 280). According to Pepper (1942/70), contextualism is also called "pragmatism." It is associated with Pierce, James, Bergson, and others.

Pepper (1942/70, pp. 248-251) underlines the implications of contextualism with regard to analysis. Whereas in formism and mechanism it is taken for granted that any object or event can be completely analyzed into its constituents, no such assumption is made in contextualism. According to contextualism only events exist and since they are totally interwoven with their context (which includes the observer), they cannot be completely analyzed. Hence one cannot get to the bottom of things. The world is bottomless and there is no ultimate nature of things because there is no-thing. There is only oneness. Since in this oneness every so-called event is interconnected with the whole cosmos, "blowing your nose is just as cosmic and ultimate as Newton's writing down his gravitational formula. The fact that his formula is much more useful to many people does not make it any more real" (Pepper 1942/70, p. 251).

The implications of contextualism for the question 'what is life?' are far-reaching. Probably the question would be considered inadequate because it is too abstract and does not arise out of concrete events in their contexts.

Living organisms must be understood in their context which is their environment, which includes other organisms as well as so-called abiotic components.

Since in contextualism change is fundamental we should use verbs instead of nouns to indicate more appropriately the acting and changing. Thus the noun 'life' is better replaced by the verb 'to live' or its gerund 'living,' which refer to an activity that occurs always in concrete situations immensely rich, complex and fluid (see, e.g., Bohm 1980).

Operation(al)ism, which can be assimilated to contextualism, also emphasizes activity in terms of operations. Meaning cannot be found in static abstraction as, for example, in formism, but must be expressed in terms of concrete operations (see Sect. 4.1.8).

## 10.2.4 Organicism

Organicism as a world hypothesis compares the whole earth and even the whole universe to an organism [see also Lovelock's (1979) Gaia hypothesis]. Everything is organized and integrated and therefore is characterized by some degree of organicity. Organicity refers to an organic whole in which every part implies every other. As a consequence, the alteration or removal of a part affects the organic whole. Seven fundamental principles are distinguished by Pepper (1942/70, p. 283). Organicism is associated with Schelling, Hegel, and modern biologists such as von Bertalanffy (1975) and Weiss (1973).

How would organicism as a world hypothesis interpret life? Probably, living organisms would be considered as regions in the universe that exhibit an extremely high degree of organcity. They are integrated into their environment, but the integration within the organisms surpasses that between organisms and environment. Looking at it this way, the apparent contradiction between the organicism of organismic biologists (Sect. 9.6) and organicism as a world hypotheses may disappear (see, e.g., Varela 1976, 1979). Organismic biologists have focused their attention at organisms in contrast to nonliving matter. Organicism as a hypothesis for the whole universe has dealt with the most inclusive whole. In this perspective nonliving and living matter of the whole universe may be seen as an analogy to nonliving parts (such as the dead wood of living tree trunks) and living parts of an organism. In both cases so-called nonliving and living components form an integrated whole. The nonliving parts are probably less integrated with the living parts than the components of the living parts with each other.

## 10.2.5  Modern Biology in Relation to the Above Four World Hypotheses

Modern biology has a heterogeneous philosophical basis in terms of the above four world hypotheses. I think that one can find traces of all four of them in various combinations. Although many biologists, especially among molecular biologists, may tend to label themselves as mechanists, their thinking may show traces of formism, or in some cases organicism or contextualism. Among typologists one may still find rather pure formists who advocate the reality of types that are characterized by essences (see, e.g., Troll 1937, 1954; Troll and Meister 1951; and many of Troll's followers). Some of the formists may tend toward mechanism, contextualism, or organicism. The organicist outlook is found among organismic biologists. The views of some of them may intergrade with mechanism, contextualism, and/or formism. Contextualism may be encountered among ecologists and again a wide overlap may occur with the other three world hypotheses. Mechanism is still widespread among molecular biologists.

At the present time it is not easy to discern trends toward one or another of the four world hypotheses. However, although mechanism is still a strong force, it seems to me that contextualism and organicism are gaining ground. Basic concepts such as "network," "multifactoriality," "dynamics," "integration," "wholeness," "continuum," "openness," "complexity," "relativity," and "complementarity" are becoming increasingly important. Dialectical approaches to biology that are related to or are part of organicism emphasize similar ideas (see, e.g., Rose 1982; Lewontin et al. 1984; Levins and Lewontin 1985). Levins and Lewontin (1985, pp. 286–288) enumerated the following principles of materialist dialectics that apply to scientific research and educational policy: historicity, universal interconnection, heterogeneity, interpenetration of opposites, and integration between levels of organization. Perhaps one might also say that a certain integration between mechanism, contextualism, and organicism is occurring. Formism, although often pronounced dead, is still with us, most often in a subconsciously hidden way. Our thinking seems to be conditioned from the cradle by the stamp of Plato and Aristotle. Only relatively few realize this so that they may liberate themselves of the strictures of formism.

The fact that the four above world hypotheses intergrade with each other indicates to me that they do not exist as fundamentally discrete doctrines. It rather seems that there is a continuum of world hypotheses from which Pepper selected four regions to which he attached the four names of formism, mechanism, contextualism, and organicism. Pepper's insistence that these four are fundamentally discrete appears to be once again the result of formist thinking which in this case claims that essentially only four forms of relatively adequate world hypotheses exist.

### 10.2.6 The Generating-substance Hypothesis

According to Pepper (1942/70, p. 92), the generating-substance hypothesis was the first self-conscious world hypothesis in European thought. It posits one (or several) substance(s) that generate the manifoldness of the universe. Thus, according to Thales who might have been the originator of this kind of world hypothesis, everything is water (Pepper 1942/70, p. 92). Anaximander's generating substance is *apeiron* or "infinite." Particular objects and qualities appeared out of this "infinite."

Pepper rejects any kind of generating-substance hypothesis because of lack of scope. I think he tends to interpret the generating substances in terms of our current scientific usage. In that sense it is, of course, impossible to generate everything from water, or air, or even a combination of substances. However, if the generating substance(s) are interpreted in a more metaphorical sense, some of them such as Anaximander's may not lack in scope as Pepper suggests.

### 10.2.7 Animism

According to animism everything in the universe is animated, i.e., endowed with a spirit. Pepper rejected animism because of inadequate precision. He presupposes, of course, implicitly that what is real has to be precise. How can such an assumption be justified? As I pointed out in the preceding chapters, both organisms and ecosystems exhibit much fluidity, fuzziness, and uniqueness. This poses severe problems with regard to the ideals of exactness and objectivity.

Pepper pointed out that animism leads to authoritarianism (Pepper 1942/70, p. 123). This may be the case to a certain extent and in this sense animism may be dangerous. Yet the fact that something is dangerous need not mean that therefore it is worthless. For example, it may be dangerous to climb a mountain, yet (s)he who succeeds reaching the top may have the most revealing and glorious view of the world. The same may apply to animism.

It is up to intuition and personal insight to distinguish between the illusory and the real. Animism in its many forms from the superficial and superstitious to the more profound is a challenge to insight and personal maturity. Neat and exact formulae or concepts will not permit us to penetrate into its secretes (if it has any).

## 10.2.8 Mysticism

In mysticism "we begin with a very impressive immediate fact, the mystic experience, a fact that is never lost sight of ... so intense that it undertakes to absorb the whole universe within it. Where it does not succeed, it denounces the unsubmissive "facts" as unreal; and since there are many of these, it spreads unreality far and wide" (Pepper 1942/70, p. 127). Hence, according to Pepper, mysticism is inadequate because of insufficient scope: many facts remain that are unsubmissive to the mystic experience.

I would like to ask Pepper how he, as someone who evidently is not a mystic and lacks the central mystical experience, can judge which facts are unsubmissive to the mystic. Pepper seems to assume that facts are given, i.e., that they are the same to a mystic and a nonmystic. He is unaware of the fact that facts may not only be theory-laden but also dependent on the state of consciousness (see Chap. 3). This means that to a mystic who is in a different state of consciousness the world may be quite different from that of an ordinary person. For example, what may be a physically or mentally painful fact to an ordinary person, may not at all be painful to a mystic. So, the question remains: how can someone who lacks the mystical experience judge how the world appears to the mystic and whether unsubmissive "facts" persist?

Mystical experience cannot be conceptualized. Hence, many mystics have preferred to remain silent or to convey their experience in poetic and paradoxical aphorisms (or koans in Zen Buddhism). Mysticism is not a hypothesis *about* the world, it is an enlightened state of being *in* the world. Hence, any philosophical representation of mysticism such as Pepper's explication of the seven categories of mysticism are bound to miss mysticism altogether. There are no categories for the mystic.

## 10.3 Truth

As mentioned already in Chap. 4 (Footnote 1), the literal meaning of "truth" is "that which is" (Bohm in Wilber 1982, p. 64). We cannot talk or write about "that which is," because as soon as we use language, which refers to concepts (abstractions), we remove ourselves from "that which is." Hence, truth cannot be communicated in words. It is in silence. When Pilatus asked Jesus 'what is thruth?', Jesus remained silent. In contrast, many philosophers and scientists have been quite verbose with regard to "truth." Evidently, they use the term 'truth' in a more restricted sense than its literal meaning indicates. In such a restricted sense quite a number of different concepts of truth can be distinguished. Thus, Pepper (1942/70) pointed out that each world hypothesis has its own concept of truth. He referred to "theories of truth" instead of "concepts of truth." Obviously, the term 'theory' is used here in the sense of meta-theory (not in the sense of a scientific theory). I shall now briefly describe what truth means in terms of Pepper's four relatively adequate world hypotheses.

*Formism*. The theory of truth in formism is called correspondence theory. According to this theory "truth consists in a similarity or correspondence between two or more things one of which is said to be true of the other" (Pepper 1942/70, p. 180). Thus pictures, maps, diagrams, sentences, formulae, or mental images may correspond to something else in nature. If they do, they are said to be true (which means that the form of the representation corresponds to that which it represents). The crux of this theory as that of formism in general is similarity or correspondence. Is there ever a total correspondence? With regard to the relation between experience and language, Bridgeman (1936, p. 23) pointed out that "nothing could be further from the truth than that there is *complete* correspondence of structure between all experience and all language, or even between any limited aspects of language and experience." If then the correspondence can be partial at best, the next question is whether the degree of partial correspondence can be high enough to provide for partial truths that in their partiality may still have significance and relevance for us (see Sect. 4.1.10).

Pepper (1942/70, p. 182) distinguished two kinds of truth in formism: historical and scientific truth. Historical truth refers to descriptions of particular (historical) events. Scientific truth entails the description of laws (or uniformities) of nature. It is thought that the form of the description corresponds to the form of the law. The question is whether there are laws in nature that have a form; and if there are such laws, the next question is how close the form of our description may approximate the form of the law.

*Mechanism*. In mechanism not seldom the correspondence theory of truth has been applied, which indicates the close connection between formism and certain versions of mechanism such as discrete mechanism that cannot dispense with identities of particulars (Pepper 1942/70, p. 222). Consolidated mechanism on the other hand has a tendency to adopt the operational theory of truth of contextualism (see below) which points to a link between mechanism and contextualism. Yet, as Pepper pointed out, mechanism has also its own theory of truth, which is called the causal-adjustment theory. According to this theory, truth "is a name for physiological attitudes which are in adjustment with the environment of the organism" (Pepper 1942/70, p. 228). For example, the sentence mentioned by Pepper (1942/70, p. 227) "That is a sharp nail" is true if my stepping on the nail causes a reaction which leads to the formulation of this sentence. Pepper claimed that in this instance nothing is implied about an identity of form between the sentence and the nail. However, he admitted that correspondence of form is implied as the connecting link between primary and secondary categories. The two kinds of categories cannot be bridged in mechanistic terms. "So the gap between the primary and the secondary categories still remains the center of the inadequacy for mechanism" (Pepper 1942/70, p. 231).

*Contextualism*. Contextualism has its own theory of truth whose modern name is operation(al)ism. Operational truth is "truth in terms of action" (Pepper 1942/70, p. 268). Pepper distinguished three versions of this theory. The first is called the "successful working" theory. Here, "truth is utility or successful functioning" (Pepper 1942/70, p. 270). For example, when a hunter whose way is interrupted by a stream manages to overcome this obstacle, his "arriving at the other bank and proceeding on his way constituted the truth of the activity" (Pepper 1942/70, p. 270). Similarily, "when a rat in a maze tries a number of blind alleys and is unsuccessful in reaching its goal, its actions are *errors*, but when it is successful in reaching its goal, it finds the *true* path" (Pepper 1942/70, p. 270). Hence, "the successful action is the true one, the unsuccessful actions are false" (Pepper 1942/70, p. 270). The question, of course, is what constitutes success. In some cases it may be clear, in others it may be debatable. In any case, it seems to be relative, depending on the situation and the context.

According to a second version of operational truth, which is called the "verified hypothesis" theory, truth is verification (in the sense of confirmation). Here, success alone does not constitute truth; truth is the relation between a hypothesis and its successful testing, which means confirmation (Pepper 1942/70, p. 272/3). However, as the first version "this theory still stresses ... that a true hypothesis gives no insight into the qualities of nature. In insists that a symbolic statement or a map or a model is no more than a tool for the control of nature. It does not mirror nature in the way supposed by the correspondence theory" (Pepper 1942/70, p. 275).

The third version of the operational theory of truth distinguished by Pepper (1942/70, p. 275) is called the "qualitative confirmation" theory. It stresses a continuity in quality between a hypothesis and the event that confirms it. "A true hypothesis, accordingly, does in its texture and quality give some insight into the texture and quality of the event it refers to for verification" (Pepper 1942/70, p. 277) (for an example see Pepper 1942/70, p. 276). This version of the operational theory of truth comes close to the correspondence theory of formism and the coherence theory of organicism (see below), yet it stresses that truth is anchored in the operation of confirmation, whereas the correspondence theory as well as the coherence theory assume that truth is a relation that is independent of the act of confirmation (Pepper 1942/70, p. 277).

*Organicism.* The theory of truth of organicism is called coherence theory. According to it, the truth of a judgement consists in its coherence (integration) with a more inclusive whole. The most inclusive whole is the absolute. In that case coherence provides for absolute truth. In the more common cases in which the judgements are integrated into less inclusive wholes, truth is of a lower level. "Thus there are degrees of truth" (Pepper 1942/70, p. 310) depending on the level of integration, i.e., coherence.

*Other world hypotheses* may also have their own notion of truth. However, to a mystic the concept (or theory) of truth vanishes altogether because (s)he has superseded discursive thought. If one insists to force a notion of truth even onto mysticism, one might say that to a mystic truth is that which is. However, not much is gained from this because that which is cannot be represented conceptually and cannot be communicated. There is no substitute for genuine and profound experience.

## 10.4 Summary

*Introduction.* Since living systems are integrated with the whole world, world hypotheses referring to the whole world are necessary for an understanding of life. World hypotheses are comprehensive world views corroborated by facts and based on a certain way of thinking. Since there are many different ways of thinking, there is a great variety of world hypotheses. Pepper (1942/70) reduced this diversity to the following seven principal world hypotheses: formism, mechanism, contextualism, organicism, generating-substance hypothesis, animism, and mysticism.

*Formism.* The central category of formism is that of (eternal) form characterized by its essence. One principal goal of formism is the reduction of observable diversity to (eternal) forms. Since external forms do not change, this view is static, while life is fluid and fuzzy. Answering the question 'what is life?' means finding the essence (eternal form) of life.

*Mechanism.* In mechanism only particulars exist. However, it is assumed that they are governed by laws. The difficult task of mechanism is the reduction of secondary qualities such as those of cells and organisms to primary qualities of size, shape, motion, weight, number, etc., and the reduction of laws concerning secondary qualities to laws holding for configurations of primary qualities. Understanding life means reducing it to the primary level and its laws.

*Contextualism.* Here the emphasis is on events in their context. Activity, change, and relation are fundamental. Law and order are not taken for granted in this pragmatic philosophy whose modern offshoot is operation(al)ism. The question 'what is life?' is not meaningful unless it relates to specific contexts.

*Organicism.* According to this world hypothesis everything is integrated with everything else. Hence, the whole earth and even the whole universe can be compared to an organism. Wholeness is the central concept. Life does not exist separately from so-called nonliving matter and therefore cannot be understood in isolation.

*Modern biology in relation to the above world hypotheses.* Formism is still found in modern biology, although rarely in a pure form as it is characteristic of certain forms of typology. Mechanism is widespread among molecular biologists, whereas organismic biologists adopt an organicist outlook. Contextualism may be encountered among ecologists. There are, of course, intergradations between and combinations of any two or more of the four principal world hypotheses. It seems that contextualism and organicism are gaining ground as there is more and more recognition of typical biological phenomena, such as dynamics, complexity, integration, and wholeness.

*Three additional world hypotheses,* according to Pepper (1942/70), are (1) the generating-substance hypothesis positing one or several substances that generate the manifoldness of the universe, (2) animism, according to which everything is endowed with a spirit, and (3) mysticism in which the mystical experience is of central importance. Mysticism should be excluded from this list, since it is not a hypothesis about the world, but refers to an enlightened state of being in the world.

*Truth.* Truth is that which is. It cannot be communicated in words. It is in silence. When philosophers and scientists speak and write about truth, they imply a limited notion of truth of which there are many. Pepper (1942/70) showed that each world hypothesis has its own concept or (meta-)theory of truth. In formism truth is the correspondence between things, such as the correspondence between a picture or concept and a natural object (correspondence theory of truth). The problem here is to what extent, if at all, such correspondence exists. The theory of truth of mechanism is called causal-adjustment theory. According to this theory truth is a formulation that results from a response (adjustment) to an environmental stimulus (cause). The correspondence theory of truth is required to relate primary and secondary categories. Contextualism has three forms of truth all of which

254

are truth in terms of action (operationism). In organicism truth is coherence (integration) with a more inclusive whole (coherence theory of truth). There are degrees of truth depending on how inclusive is the whole. Other world hypotheses may also have their own concept of truth. However, to a mystic the concept of truth vanishes because (s)he has superseded discursive thought. If one insists to force a notion of truth onto mysticism, one might say that to a mystic truth is that which is.

# Epilogue

"Everything that is thought and expressed in words is one-sided, only half the truth; it lacks totality, completeness, unity" (Hermann Hesse: Siddhartha 1957, p. 115).

## On the Importance of Living

In writing a book like this it seems improbable to avoid mistakes. I hope I have kept their number and magnitude to a minimum. Furthermore, it is impossible to avoid omissions of relevant ideas, topics, points of views, etc. Although I have tried to present a great diversity of topics and points of view, completeness is unattainable. In the unavoidable process of selection, much is left out and thus a one-sided, (over)simplified and even distorted picture results. I hope again that the (over)simplification and distortion have been kept to a minimum, but the reader must be warned.

Apart from one-sidedness and distortion that result from the selection of certain statements and ways of phrasing them, any verbal communication that uses concepts implies further one-sidedness that is often overlooked: as pointed out in the Chapter on *Concepts and Classification*, concepts are the result of abstraction which is selective attention. Thus, any conceptual representation of life and the world is restrictive. Life is always infinitely more than can be said or written about it. To understand it profoundly, it has to be lived in such a way that living and understanding become one: hence, the importance of living ...

Fromm (1976), in a book entitled *To have or to be?*, distinguished two modes of existence: the having mode and the being mode. Most of us live predominantly in the having mode. We want to have more and more money, security, happiness, knowledge, wisdom, etc. It would seem that there is a great difference between seeking material goods and knowledge or wisdom. Nonetheless, all seeking is an expression of the having mode. Living according to this mode may allow us, under certain conditions, to acquire material goods and knowledge. Profound happiness and wisdom are, however, incompatible with this mode of living; they require living in the being mode.

256

The being mode refers to experience in the most encompassing way including all the subjective nuances. Since such experience is infinite (i.e., open toward the whole universe), it cannot be described and therefore the being mode cannot be defined or characterized: it must be experienced. This is the challenge and joy of living.

In his *Lectures on Zen Buddhism*, D.T. Suzuki (in Fromm et al. 1960) contrasts a poem by Tennyson with a haiku (japanese poem) by Bashu to illustrate the difference between West and East. Fromm (1976) reproduces these two poems to indicate the difference between the having and being modes of existence. Tennyson's verse is as follows:

> Flower in the crannied wall,
> I pluck you out of the crannies,
> I hold you here, root and all, in my hand,
> Little flower – but *if* I could understand
> What you are, root and all, and all in all,
> I should know what God and man is

Bashu's haiku (translated into English) runs like this:

> When I look carefully
> I see the *nazuna* blooming
> By the hedge!

As pointed out by Suzuki and Fromm, Tennyson "needs to possess the flower in order to understand people and nature, and by his *having* it (plucking it), the flower is destroyed" (Fromm 1976, p. 17). In contrast Basho just looks carefully at the flower (nazuna) and "feels all the mystery ... that goes deep into the source of all existence" (Suzuki in Fromm et al. 1960). He does not separate and analyze the flower; in his nonintrusive, nonmanipulative looking or seeing he can let it live, let it be ...

# References

Achinstein P (1983) The nature of explanation. Oxford Univ Press, New York

Ames MM (1983) How should we think about what we see in a museum of anthropology? Trans R Soc Can (Ser IV) 21:93−101

Arber A (1946) Goethe's botany. The metamorphosis of plants (1790) and Tobler's Ode to Nature (1782) with an introduction and translation by Agnes Arber. Chron Bot 10:66−124

Arber A (1950) The natural philosophy of plant form. Cambridge Univ Press, Cambridge (reprinted 1970 by Hafner, Darien, Conn. USA)

Arber A (1954) The mind and the eye. Cambridge Univ Press, Cambridge (paperback edn, 1964)

Arber A (1957) The manifold and the One. Theosophical, Wheaton, Ill.

Atlan H (1979) Entre le cristal et la fumée. Essai sur l'organisation du vivant. Seuil, Paris

Axelrod R (1984) The evolution of cooperation. Basic Books, New York

Axelrod R, Hamilton WD (1981) The evolution of cooperation. Science 211:1390−1396

Ayala FJ (1968) Biology as an autonomous science. Am Nat 56:207−227

Ayala FJ (1974a) Introduction to studies in the philosophy of biology. In: Ayala FJ, Dobzhansky T (eds) Studies in the philosophy of biology. Reduction and related problems. Univ California Press, Berkeley Los Angeles, pp VII−XVI

Ayala FJ (1974b) The concept of biological progress. In: Ayala FJ, Dobzhansky T (eds) Studies in the philosophy of biology. Reduction and related problems. Univ California Press, Berkeley Los Angeles, pp 339−356

Ayala FJ, Dobzhansky T (eds) (1974) Studies in the philosophy of biology. Reduction and related problems. Univ California Press, Berkeley Los Angeles

Babb LA (1976) Thaipusam in Singapore: religious individualism in a hierarchical culture. Sociology working paper no 49. Dept Sociobiol, Univ Singapore, Chopmen Enterprises, Singapore

Backster C (1968) Evidence of a primary perception in plant life. Int J Parapsychol 10:329−348

Barash DP (1973) The ecologist as Zen master. Am Midl Nat 89:214−217

Barash DP (1977) Sociobiology and behavior. Elsevier, New York

Barash DP (1979) The whisperings within: evolution and the origin of human nature. Harper and Row, New York

Barber B (1961) Resistance by scientists to scientific discovery. Science 134:596−602

Barlow PW (1982) The plant forms cells, not cells the plant: the origin of de Bary's aphorism. Ann Bot 49:269−271

258

Barlow PW, Carr DJ (eds) (1984) Positional controls in plant development. Cambridge Univ Press, Cambridge

Bateson G (1972) Steps to an ecology of mind. Ballantine Books, New York

Bateson G (1979) Mind and nature. A necessary unity. Dutton, New York (paperback by Bantam Books, 1980)

Beatty J (1980) What's wrong with the received view of evolutionary theory? Philos Sci Assoc 2:397–426

Beatty J (1982) Classes and cladists. Syst Zool 31:25–34

Beckner M (1959) The biological way of thought. Columbia Univ Press, New York (2nd edn Univ California Press, 1968)

Beckner J (1969) Function and teleology. J Hist Biol 2:151–164

Bernier R, Pirlot P (1977) Organe et fonction. Essai de biophilosophie. Maloine-Doin, Paris, Edisem, St. Hyacinthe, Quebec

Bergson HL (1889) Essai sur les donnés immédiates de la conscience. Alcan, Paris (translated into English as "Time and free will: an essay on the immediate data of consciousness". Macmillan, New York, 1910)

Bergson HL (1900) Le rire. Alcan, Paris (translated into English as "Laughter. An essay on the meaning of the comic". MacMillan, New York, 1911)

Bergson HL (1907) L'évolution créatrice. Alcan, Paris (translated into English as "Creative Evolution". Modern Library, New York, 1944)

Berofsky B (ed) (1966) Free will and determinism. Harper and Row, New York

Bertalanffy L von (1952) Problems of life. Watts, London (Harper Torchbook edn, 1960)

Bertalanffy L von (1965) Zur Geschichte theoretischer Modelle in der Biologie. Stud Gen 18:290–298

Bertalanffy L von (1967) Robots, men and minds. Braziller, New York

Bertalanffy L von (1968) General system theory. Braziller, New York

Bertalanffy L von (1975) In: Taschdjian E (ed) Perspectives on General System Theory. Braziller, New York

Bieniek ME, Millington WF (1967) Differentiation of lateral shoots as thorns in *Ulex europaeus*. Am J Bot 54:61–70

Bindra D (1976) A theory of intelligent behavior. Wiley Interscience, New York

Birch C, Cobb JB Jr (1981) The liberation of life -- from the cell to the community. Cambridge Univ Press, Cambridge

Birx HJ (1984) Theories of evolution. Thomas, Springfield, Ill. USA

Bischof N (1977) Verstehen und Erklären in der Wissenschaft vom Menschen. In: Lohmann M (ed) Wohin führt die Biologie? Ein interdisziplinäres Kolloquium. Deutscher Taschenbuch Verlag, München

Black S (1972) The nature of living things. Seeker and Warburg, London

Blandino G (1969) Theories on the nature of life. Philosophical Library, New York

Boden MA (1979) The computational metaphor in psychology. In: Bolten N (ed) Philosophical problems in psychology. Methuen, London

Boden MA (1981) Minds and mechanisms: a philosophical psychology and computational models. Cornell Univ Press, Ithaca, New York

Bohm D (1971) Fragmentation in science and society. In: Fuller W (ed) The social impact of modern biology. Routledge and Kegan Paul, London

Bohm (1973) Human nature as the product of our mental models. In: Benthall J (ed) The limits of human nature. Inst Contemporary Arts, London (Dutton paperback, 1974)

Bohm (1980) Wholeness and the implicate order. Routledge and Kegan Paul, London

Bookstein FL (1978) The measurement of biological shape and shape change. Springer, Berlin Heidelberg New York (Lect Notes Biomath 24)

Boucher DH (1981) The gospel according to sociobiology. Perspect Biol Med 25:63–65

Boucher DH (ed) (1985) The biology of mutualism. Croom Helm, London

Boucher DH, James S, Keeler KH (1982) The ecology of mutualism. Annu Rev Ecol Syst 13:315–347

Bradie M, Gromko M (1981) The status of the principle of natural selection. Nature Syst 3:3–12

Brand M (1979) Causality. In: Asquith PD, Kyburg HE Jr (eds) Current research in philosophy of science. Philos Sci Assoc, East Lansing, Michigan, pp 252–281

Bradon RN (1981) Biological teleology: questions and explanations. Stud Hist Philos Sci 12:91–105

Brandon RN, Burian RM (eds) (1984) Genes, organisms, populations. MIT Press, Cambridge

Brandt LW (1982) Psychologists caught: a psycho-logic of psychology. Univ Toronto Press, Toronto

Bretschneider J, Kaiser H, Plesse W, Schellhorn M (1982) Philosophische Aspekte der Biologie. Fischer, Jena

Bridgman PW (1927) The logic of modern physics. Macmillan, New York

Bridgman PW (1936) The nature of physical theory. Princeton Univ Press, Princeton (also Dover paperback)

Bronowski J (1966) The logic of the mind. Am Sci 54:1–14

Bronowski J (1973) The ascent of man. Little and Brown, Boston

Brooks DR, Wiley EO (1984) Evolution as an entropic phenomenon. In: Pollard JW (ed) Evolutionary theory. Paths into the future. Wiley, New York, pp 141–172

Browder LW (1980) Developmental biology. Saunders College, Philadelphia

Brown HI (1977) Perception, theory and commitment: the new philosophy of science. Precedent, Chicago

Buckley P, Peat FD (eds) (1979) A question of physics. Conversations in physics and biology. Univ Toronto Press, Toronto

Bunge M (1967) Scientific research, vol 1. Springer, Berlin Heidelberg New York

Bunge M (1968) Conjunction, succession, determination and causation. Int J Theor Physics 1:299–315

Bunge M (1973) Method, model and matter. Reidel, Dordrecht

Bunge M (1977) The GST challenge to the classical philosophies of science. Int J Gen Syst 4:29–37

Bunge M (1979a) Causality and modern science, 3rd revised edn. Dover, New York

Bunge M (1979b) A world of systems. Treatise on basic philosophy, vol 4. Reidel, Dordrecht

Bunge M (1980) The mind-body problem – a psychobiological approach. Pergamon, London

Bünning E (1949) Theoretische Grundfragen der Physiologie, 2nd edn. Piscator, Stuttgart

Bünning E (1977) Fifty years of research in the wake of Wilhelm Pfeffer. Annu Rev Plant Physiol 28:1–22

Buss LW (1983) Somatic variation and evolution. Paleobiology 9:12–16

Butts RE (1976) The hypothetico-deductive model of scientific theories: a sympathetic disclaimer. In: Shea WR (ed) Basic issues in the philosophy of science. Science History, Neale Watson, New York (Can Contemporary Philosophy Ser, pp 36–57)

Campbell DT (1974) Evolutionary epistemology. In: Schilpp PA (ed) The philosophy of KR Popper. Open Court, La Salle, Ill.

260

Canfield JV (ed) (1966) Purpose in nature. Prentice Hall, Englewood Cliffs, NJ

Canguilhem G (1975) La connaissance de la vie. Vrin, Paris

Cannon WB (1936) The wisdom of the body. Norton, New York

Caplan A (ed) (1978) The sociobiology debate. Harper and Row, New York

Caplan A (1981) Back to class: a note on the ontology of species. Philos Sci 48:130–140

Capra F (1975) The Tao of physics. An exploration of the parallels between modern physics and Eastern mysticism. Shambhala, Berkeley

Capra F (1982) The turning point: science, society and the rising culture. Simon and Schuster, New York

Caratini R (1972) Encyclopédie thématique universelle. Bordas, Paris

Carnap R (1936/7) Testability and meaning. Philos Sci 3:419–471, and 4:1–40

Castaneda C (1972) A separate reality. Further conversations with Don Juan. Pocket Books, New York

Castaneda C (1974) Journey to Ixtlan. The lessons of Don Juan. Pocket Books, New York

Chant C, Fauvel J (eds) (1980) Darwin to Einstein: historical studies on science and belief. Longman, in association with The Open Univ Press, Harlow, Essex

Cohen IB (1985) Revolutions in science. Belknap Press of Harvard Univ Press, Cambridge, MA London

Cohen LJ, Hesse M (eds) (1980) Applications of inductive logic. Proc Conf Queen's College, Oxford. Clarendon, Oxford

Coley NG, Hall VMD (eds) (1980) Darwin to Einstein: primary sources on science and belief. Longman, in association with the Open Univ Press, Harlow, Essex

Collingwood RG (1940) An essay on metaphysics. Oxford Univ Press, New York

Cousins N (1979) Anatomy of an illness as perceived by the patient. Reflections on healing and regeneration. Bantam Books, Norton, Toronto

Cracraft J (1983) Cladistic analysis and vicariance biogeography. Am Sci 71:273–281

Crick F (1966) Of molecules and men. Univ Washington Press, Seattle

Crick F (1981) Life itself: its origin and nature. Simon and Schuster, New York

Croizat L (1960) Principia botanica, 2 vols. Caracas (published by the author)

Croizat L (1962) Space, time, form: the biological synthesis. Caracas (published by the author)

Crook JH (1983) On attributing consciousness to animals. Nature 303:11–14

Culler J (1982) On deconstruction. Theory and criticism after structuralism. Cornell Univ Press, Ithaca, New York

Cusset G (1982) The conceptual bases of plant morphology. Acta Biotheor 31A:8–86

Cusset G (1985) Tige et feuille de Dicotylédones: aspects morphogénétiques. Bull Assoc Dev Méth Théor Biol (Coif sur Yvette) (in press)

Cusset G (1986) La morphogenèse du limbe des Dicotylédones. Can J Bot (in press)

Da Liu (1974) Taoist health exercise book. Links Books, New York

Davenport R (1979) An outline of animal development. Addison-Wesley, Reading, MA

Davies P (1983) God and the new physics. Penguin Books, New York

Davis M, Lane E (1978) Rainbows of life: the promise of Kirlian photography. Harper and Row, New York

Davis PH, Heywood VH (1963) Principles of Angiosperm taxonomy. Oliver and Boyd, Edinburgh

De Duve C (1984) A guided tour of the living cell, vol 1. Scientific American Library, Freeman, New York

De Robertis EDP, De Robertis EMF (eds) (1980) Cell and molecular biology, 7th edn. Saunders College, Philadelphia

Dentor M (1984) Evolution. A theory in crisis. Bernet, London

261

Deshpande PY (1978) The authentic Yoga. Patanjali's yoga sutras. Rider, London
Dickinson TA (1978) Epiphylly in angiosperms. Bot Rev 44:181–232
Dickinson TA, Sattler R (1974) Development of the epiphyllous inflorescence of *Phyllonoma integerrima* (Turcz.) Loes: implications for comparative morphology. Bot J Linn Soc 69:1–13
Dickinson TA, Sattler R (1975) Development of the epiphyllous inflorescence of *Helwingia japonica* (Helwingiaceae). Am J Bot 62:962–973
Dickison WC, White RA (eds) (1984) Contemporary problems in plant anatomy. Academic Press, New York
Dilworth C (1981) Scientific progress. A study concerning the nature of the relation between successive scientific theories. Synthese Library, Reidel, Dordrecht
Ditfurth H von (1979) An der Grenze zwischen Geist und Biologie. Der Spiegel 40:249, 252, 254, 256
Dobzhansky T, Ayala FJ, Stebbins GL, Valentine JW (1977) Evolution. Freeman, San Francisco
Dörner D (1975) Wie Menschen eine Welt verbessern wollten und sie dabei zerstörten. Bild Wiss 48–53
Dray WH (1964) Philosophy of history. Prentice Hall, Englewood Cliffs, NJ
Driesch HAE (1908) Science and philosophy of the organism. Adam Charles Black, London
Dubois D, Prade H (1980) Fuzzy sets and systems. Academic Press, New York
Dubos R (1961) The dreams of reason. Columbia Univ Press, New York
Dubos R (1981) Celebration of life. McGraw Hill, New York
Dullemeijer P (1974) Concepts and approaches in animal morphology. Van Gorcum, Assen, Netherlands
Dullemeijer P (1980) Functional morphology and evolutionary biology. Acta Biotheor 29:151–250
Dupre J (1981) Natural kinds and biological taxa. Philos Rev 90:66–90
Eames AJ (1961) Morphology of angiosperms. McGraw Hill, New York
Edge D (1983) Is there too much sociology of science? Isis 74:250–256
Eigen M, Winkler R (1981) Laws of the game. How the principles of nature govern chance. Knopf, New York
Eldredge N, Cracraft J (1980) Phylogenetic patterns and the evolutionary process. Method and theory in comparative biology. Columbia Univ Press, New York
Eldredge N, Gould SJ (1972) Punctuated equilibria: an alternative to phyletic gradualism. In: Schopf TJM (ed) Models in paleobiology. Freeman and Cooper, San Francisco, pp 82–115
Elsasser WM (1975) Chief abstractions of biology. American Elsevier, New York
Elsasser WM (1981) Principles of a new biological theory: a summary. J Theor Biol 89:131–150
Engel G (1977) The need for a new medical model: a challenge for biomedicine. Science 196:129–136
Eyde RH (1975) The foliar theory of the flower. Am Sci 63:430–437
Feldenkrais M (1972/77) Awareness through movement. Harper and Row, New York
Ferguson A (1976) Can evolutionary theory predict? Am Nat 110:1101–1104
Ferguson M (1980) The aquarian conspiracy. Tarcher, Los Angeles
Ferry G (1984) The understanding of animals. Basil Blackwell and New Scientist, Oxford
Feyerabend PK (1975) Against method: an outline of an anarchistic theory of knowledge. In: Radner M, Winokur S (eds) Minn Stud Philos Sci 4:17–130. Univ Minnesota Press, Minneapolis (Verso edn, paperback, 1978)

Feyerabend PK (1981a) Realism, rationalism and scientific method. Philosophical papers, vol 1. Cambridge Univ Press, Cambridge

Feyerabend PK (1981b) Problems of empiricism. Philosophical papers, vol 2. Cambridge Univ Press, Cambridge

Feyerabend PK (1982) Science in a free society. Verso edn, NLB, London (Originally published in 1978 by NLB, London)

Fleck L (1979) Genesis and development of a scientific fact. Univ Chicago Press, Chicago

Fleck L (1981) On the question of the foundation of medical knowledge. (Translated by Trenn TJ). J Med Philos 6:237–256

Fromm E (1956) The art of loving. Harper and Row, New York

Fromm E (1976) To have or to be? Harper and Row, New York (World Perspectives, vol 50, ed by RN Anshen)

Fromm E, Suzuki DT, De Martino R (1960) Zen Buddhism and psychoanalysis. Harper and Row, New York

Gabor D (1964) Inventing the future. Penguin Books, Harmondsworth, England

Galston AW, Slayman CL (1979) The not-so-secret life of plants. Am Sci 67:337–344

Ghiselin MT (1974) A radical solution to the species problem. Syst Zool 23:523–536

Giere RN (1979) Understanding scientific reasoning. Holt, Rinehart and Winston, New York

Gilmour JSL (1940) Taxonomy and philosophy. In: Huxley J (ed) The new systematics. Clarendon, Oxford, pp 461–474

Gilmour JSL (1961) Taxonomy. In: Macleod AM, Cobley LS (eds) Contemporary botanical thought. Oliver and Boyd, Edinburgh, pp 27–45

Gilson E (1971) D'Aristote à Darwin et retour. Vrin, Paris

Gleason HA (1961) An introduction to descriptive linguistics. Holt, Rinehart and Winston, New York

Goethe JW von (1790) Versuch die Metamorphose der Pflanzen zu erklären. Ettinger, Gotha

Good R (1981) The philosophy of evolution. Dovecote, Stanbridge, Wimborne, Dorset

Goodwin BC, Holder N, Wylie CC (eds) (1983) Development and evolution. Cambridge Univ Press, Cambridge

Gorelik G (1975) Principal ideas of Bogdonov's 'tektology': the universal science of organization. Gen Syst 20:3–13

Gottlieb LD (1984) Genetics and morphological evolution in plants. Am Nat 123: 681–709

Goudge TA (1961) The ascent of life. A philosophical study of the theory of evolution. Univ Toronto Press, Toronto

Gould SJ (1977) Ontogeny and phylogeny. Harvard Univ Press, Cambridge, MA

Gould SJ (1980) Is a new and general theory of evolution emerging? Paleobiology 6: 119–130

Gould SJ (1981) The mismeasure of man. Norton, New York

Gould SJ, Eldredge N (1977) Punctuated equilibria: the tempo and mode of evolution reconsidered. Paleobiology 3:115–151

Gould SJ, Lewontin RC (1979) The spandrels of San Marco and the Panglossian paradigm: a critique of the adaptationist program. Proc R Soc Lond B Biol Sci 205: 581–598

Grant V (1981) Plant speciation, 2nd edn. Columbia Univ Press, New York

Gray W, Rizzo ND (eds) (1973) Unity through diversity. A Festschrift for Ludwig von Bertalanffy. In 2 parts. Gordon and Breach Science, New York

Greene JC (1971) The Kuhnian paradigm and the Darwinian revolution in natural history. In: Roller DHD (ed) Perspectives in the history of science and technology. Univ Oklahoma Press, Norman, Okla

Greene JC (1981) Science, ideology, and world view: essay in the history of evolutionary ideas. Univ California Press, Berkeley

Gregory RL, Gombrich EH (eds) (1973) Illusion in nature and art. Duckworth, London

Grene M (1965) Approaches to philosophical biology. Basic Books, New York London

Grene M (1974) The understanding of nature. Reidel, Dordrecht

Grene M (1980) A note on Simberloff's 'succession of paradigm in ecology'. Synthese 43:40–45

Grene M (ed) (1983) Dimensions of Darwinism. Cambridge Univ Press, Cambridge

Grene M, Mendelsohn E (eds) (1976) Topics in the philosophy of biology. Reidel, Dordrecht (Boston Stud Philos Sci 27)

Grene M, Burian R (eds) (1983) Philosophy of biology in the philosophy curriculum. A publication developed at the 1982 Summer Institute of Philosophy of Biology. Council of Philosophical Studies, San Francisco

Griffin DR (1981) The question of animal awareness. Evolutionary continuity of mental experience. Revised and enlarged edn. Rockefeller, New York

Griffin DR (ed) (1982) Animal mind – human mind. Report of the Dahlem Workshop, Berlin 1981. Springer, Berlin Heidelberg New York

Griffin DR (1984) Animal thinking. Do animals have conscious awareness– Am Sci 72: 456–464

Grosvernor D, Grosvernor G (1966) Ceylon. Nat Geogr Magazine 129:447–497

Guédès M (1979) Morphology of seed-plants. Cramer, Vaduz

Gunning BES, Overall RL (1983) Plasmodesmata and cell-to-cell transport in plants. Bioscience 33:260–265

Gutting G (ed) (1980) Paradigms and revolutions: appraisals and applications of Thomas Kuhn's philosophy of science. Univ Notre Dame Press, Notre Dame

Habermas J (1970) Technik und Wissenschaft als Ideologie. Suhrkamp, Frankfurt

Hacking I (ed) (1981) Scientific revolutions. Oxford Univ Press, Oxford

Hagemann W (1982) Vergleichende Morphologie und Anatomie – Organismus und Zelle, ist eine Synthese möglich? Ber Dtsch Bot Ges 95:45–56

Haken H (1977) Synergetics. Springer, Berlin Heidelberg New York

Hallé F, Oldeman RAA, Tomlinson PB (1978) Tropical trees and forests. An architectural analysis. Springer, Berlin Heidelberg New York

Hammen L van der (1983) Unfoldment and manifestation: the natural philosophy of evolution. Acta Biotheor 32:179–194

Hanson NR (1958) Patterns of discovery. An inquiry into the conceptual foundations of science. Cambridge Univ Press, Cambridge

Haraway DJ (1976) Crystals, fabrics, and fields. Metaphors of organicism in twentieth-century developmental biology. Yale Univ Press, New Haven London

Harris H (1981) Rationality in science. In: Heath AF (ed) Scientific explanation. Clarendon Press, Oxford

Hartmann M (1948) Die philosophischen Grundlagen der Naturwissenschaften. Erkenntnistheorie und Methodologie. Fischer, Jena

Hasler A, Scholz A (1983) Olfactory imprinting and homing in salmon. Springer, Berlin Heidelberg New York

Hassenstein B (1954) Abbildende Begriffe. Verhandl Dtsch Zool Ges Tübingen, pp 197–202

Hassenstein B (1971) Injunktion. In: Ritter J (ed) Historisches Wörterbuch der Philosophie, vol 4, pp 367–368. Schwabe, Basel

Hassenstein B (1978) Wie viele Körner ergeben einen Haufen? In: Preisl A, Mohler A (eds) Vol 1: Der Mensch und seine Sprache. Propyläen Verlag, Berlin, pp 219–242

Hayward WJ (1984) Perceiving ordinary magic. Science and intuitive wisdom. New Science Library, Shambhala, Boulder, Co.

Heads M (1984) Principia Botanica: Croizat's contribution to botany. In: Craw RC, Gibbs GW (eds) Croizat's Panbiogeography and Principia Botanica: search for a novel biological synthesis. Tuatara 27:26−48

Heaton JM (1979) Theory of psychotherapy. In: Bolten N (ed) Philosophical problems in psychology. Methuen, London, pp 176−196

Heidcamp WH (ed) (1978) The nature of life. Univ Park Press, Baltimore

Heisenberg W (1962) Physics and philosophy. The revolution in modern science. Harper and Row (Harper Torchbooks), New York

Hempel CG (1965) Aspects of scientific explanation. The Free Press, New York

Hempel CG (1966) Philosophy of natural science. Prentice Hall, Englewood Cliffs, NJ

Hertel R (1980) Prediction and control: the only aim of biological research? Riv Biol (Perugia) 73:489−506

Hesse H (1957) Siddhartha. Translated by Hilda Rosner. New Directions (paperback no 65), New York

Hesse MB (1970) Is there an independent observation language? In: Colodny RG (ed) The nature and function of scientific theories. Essays in contemporary science and philosophy. Univ Pittsburgh Press, Pittsburgh

Hesse MB (1974) The structure of scientific inference. Univ California Press, Berkeley

Hilborn R, Stearns SC (1982) On inference in ecology and evolutionary biology: the problem of multiple causes. Acta Biotheor 31:145−164

Hirsch A (1975) Is there a need for a new journal of structural botany? Plant Sci Bull 21:53−54

Hitching F (1982) The neck of the giraffe. Where Darwin went wrong. Ticknor and Fields, New Haven

Ho MW, Saunders PT (1979) Beyond neo-Darwinism − an epigenetic approach to evolution. J Theor Biol 78:573−591

Ho MW, Saunders PT (eds) (1984) Beyond neo-Darwinism. An introduction to the new evolutionary paradigm. Academic Press, New York

Hofstadter DR (1979) Gödel, Escher, Bach. Basic Books, New York

Hofstadter DR, Dennett DC (1981) The mind's I. Harvester Basic Books, New York

Horowitz KA, Lewis DC, Gasteiger EL (1975) Plant "primary perception": electrophysiological unresponsiveness to brine shrimp killing. Science 189:478−480

Howard RA (1974) The stem-node-leaf continuum of the Dicotyledoneae. J Arnold Arbor Univ 55:125−181

Hull DL (1965) The effect of essentialism on taxonomy. Two thousand years of stasis. Br J Philos Sci 15:314−26 and 16:1−18

Hull DL (1968) The operational imperative: sense and nonsense in operationism. Syst Zool 17:438−457

Hull DL (1974) Philosophy of biological science. Prentice Hall, Englewood Cliffs, NJ

Hull DL (1976) Are species really individuals. Syst Zool 25:174−191

Hull DL (1980) Individuality and selection. Annu Rev Ecol Syst 11:311−332

Hull DL (1981a) Historical narratives and integrating explanations. In: Sumner LW (ed) Pragmatism and purpose. Univ Toronto Press, Toronto, pp 172−188

Hull DL (1981b) Kitts and Kitts and Caplan on species. Philos Sci 48:141−152

Hull DL (1983) Review of "M. Ruse: Darwinism defended: a guide to evolution controversies". Isis 74:106−107

Huxley A (1963) Literature and science. Chatto and Windus, London

Huxley JS (1942) Evolution: the modern synthesis. Harper, New York

Huxley JS (1953) Evolution in action. Harper, New York

I Ching (Book of Changes). The Richard Wilhelm translation rendered into English by Baynes CF. Foreword by CG Jung. Princeton Univ Press, Princeton, 3rd edn, 1967

Illich I (1976) Medical nemesis: the expropriation of health. Pantheon (Div of Random House), New York

Izutsu T (1967) The Absolute and the perfect man in Taoism. Eranos Jahrb 36:379–440

Izutsu T (1971) The philosophy of Zen. In: Klibansky R (ed) Contemporary philosophy. A survey. La Nuova Italia Editrice, Firenze

Izutsu T (1974) The philosophical problem of articulation in Zen Buddhism. Rev Int Philos 107–108:165–183

Jacob F (1970) La logique du vivant. Gallimard, Paris (English translation: The logic of life. A history of heredity. Pantheon, New York, 1973)

Jeuken M (1968) A note on models and explanation in biology. Acta Biotheor 18:284–290

Jeuken M (1975) The biological and philosophical definitions of life. Acta Biotheor 24:14–21

Jeune B (1981) Modèle empirique du développement des feuilles des Dicotylédones. Bull Mus Natl Hist Nat Paris, 4e sér, 3 (sect B):433–459

Jevons FR (1973) Science observed. Science as a social and intellectual activity. Allen and Unwin, London

Johnson PL (ed) (1977) An ecosystem paradigm for ecology. Oak Ridge Assoc Univ

Jonas H (1966) The phenomenon of life. Harper and Row, New York

Jong K (1978) Phyllomorphic organisation in rosulate *Streptocarpus*. Notes R Bot Garden Edinburgh 36:369–396 (BL Burtt Festschrift)

Jong K, Burtt BL (1975) The evolution of morphological novelty exemplified in the growth patterns of some Gesneriaceae. New Phytol 75:297–311

Jordan P (1941) Die Physik und das Geheimnis des organischen Lebens. Vieweg, Braunschweig

Jordan P (1972) Wie frei sind wir? Naturgesetz und Zufall. Fromm, Osnabrück

Jung CG (1967) Foreword to the I Ching (Book of Changes). Princeton Univ Press, Princeton

Kant I (1781) Kritik der reinen Vernunft. Hartknoch, Riga (Suhrkamp, 1977) (translated into English by Müller FM: Critique of pure reason. Doubleday, Garden City, New York, 1966)

Kapleau P (1979) Zen: dawn in the West. Anchor, Doubleday, Garden City, New York

Kaussmann B (1955) Histogenetische Untersuchungen zum Flachsproßproblem. Bot Stud 3:1–136

Keller EF (1983) A feeling for the organism: the life and work of Barbara McClintock. Freeman, San Francisco

Kimura M (1979) The neutral theory of molecular evolution. Sci Am 241:98–130

Kitts DB, Kitts DJ (1979) Biological species as natural kinds. Philos Sci 46:613–622

Kleiber M (1961) The fire of life. An introduction to animal energetics. Wiley, New York

Knox RB, Considine JA (1982) Commentary: deterministic and probabilistic approaches to plant development. Acta Biotheor 31A:112–117

Kochanski Z (1973) Conditions and limitations of prediction-making in biology. Philos Sci 40:29–51

Koestler A (1967) The ghost in the machine. Hutchinson, London (Paperback by Pan Books, 1970)

Koestler A, Smythies JR (eds) (1969) The Alpach Symposium 1968. Beyond reductionism: new perspectives in the life sciences. Hutchinson, London

Kohler J, Kohler MA (1979) Healing miracles from macrobiotics. Parker, West Nyack, New York

266

Kozlovsky DG (1974) An ecological and evolutionary ethic. Prentice-Hall, Englewood Cliffs, New York

Krishnamurti J (1970) Talks and dialogues. Avon Books, New York

Krishnamurti J (1985) Mind without measure. Krishnamurti Foundation, Madras, India

Krishnamurti J, Bohm D (1985) The ending of time. 13 dialogues between Krishnamurti and David Bohm. Harper and Row, New York

Kropotkin P (1902) Mutual aid: a factor in evolution. Heinemann, London (1972 edn with an introduction by Avrich P. Porter Sargent, Boston)

Kuhn TS (1970) The structure of scientific revolutions, 2nd edn. Univ Chicago Press, Chicago

Kuhn TS (1977) The essential tension: selected studies in scientific tradition and change. Univ Chicago Press, Chicago

Kurzweil R (1985) What is artificial intelligence anyway? Am Sci 73:258–264

Kushi M (1978) Natural healing through macrobiotics. Japan, Tokyo

Kushi M (1981) The macrobiotic approach to cancer. Towards preventing and controlling cancer with diet and lifestyle. Avery, Wayne, NJ

Kushi M (1982) Cancer and heart disease: the macrobiotic approach to degenerative disorders. Japan, Tokyo

Lakatos I (1968) Criticism and the methodology of scientific research programmes. Proc Arist Soc 69:149–189

Lakatos I (1976) Falsification and the methodology of scientific research programmes. In: Harding SG (ed) Can theories be refuted? Essays on the Duhem-Quine thesis. Synthese Library 81. Reidel, Dordrecht, Holland, pp 205–259

Lakatos I (1978) Philosophical papers, 2 vols (ed by Worrall J, Currie G). Cambridge Univ Press, Cambridge

Lakatos I, Musgrave A (eds) (1970) Criticism and growth of knowledge. Cambridge Univ Press, Cambridge

Langer SK (1964) Philosophical sketches. Mentor Book, by arrangement with Hopkins Press, Baltimore

Laotse (1948) The wisdom of Laotse. Translated, edited, and with an introduction and notes by Lin Yutang. Modern Library, Random House, New York

Lao Tsu, Tao Te Ching Translated by Gia-Fu Feng and Jane English (1972) Vintage Books, New York

Laszlo E (1972a) The systems view of the world. A natural philosophy of new developments in the sciences. Braziller, New York

Laszlo E (ed) (1972b) The relevance of General Systems Theory. Papers presented to Ludwig von Bertalanffy on his 70th birthday. Braziller, New York

Laszlo E (1973) A general systems model of the evolution of science. Scientia (Milan) 107:379–395

Laszlo E (1974) A strategy for the future. The systems approach to world order. Braziller, New York

Leggett WC (1977) The ecology of fish migrations. Annu Rev Ecol Syst 8:285–308

Leiss W (1972) The domination of nature. Braziller, New York

LeShan L (1976) Alternate realities. The search for the full human being. Evans, New York

LeShan L, Margenau H (1982) Einstein's space and van Gogh's sky. Physical reality and beyond. Macmillan, New York

Levins R (1968) Evolution in changing environments. Some theoretical explorations. Princeton Univ Press, Princeton, NJ

Levins R, Lewontin RC (1980) Dialectics and reductionism in ecology. Synthese 43:47–78

267

Levins R, Lewontin RC (1985) The dialectical biologist. Harvard Univ Press, Cambridge, MA

Lewin R (1983) Predators and hurricanes change ecology. Science 221:737–740

Lewis DH (1973) Concepts in fungal nutrition and the origin of biotrophy. Biol Rev 48:261–278

Lewis ER (1977) Network models in population biology. Springer, Berlin Heidelberg New York

Lewontin RC (1966) Is nature probable or capricious. Bioscience 16:25–27

Lewontin RC (1972) Book review: Creed R (ed) Ecological genetics and evolution. Nature 236:181–182

Lewontin RC (1981) Sleight of hand. Review of Lumsden CJ, Wilson EO (eds) Genes, mind, and culture. The co-evolutionary process. Sciences (NY) 21:23–26

Lewontin RC, Levins R (1976) The problem of Lysenkoism. In: Rose H, Rose S (eds) The radicalisation of science. Macmillan, London, pp 32–64

Lewontin RC, Rose S, Kamin LJ (1984) Biology, ideology, and human nature. Not in our genes. Pantheon Books, New York

Lin Yutang (1938) The importance of living. Heinemann, London

Lindenmayer A (1978) Algorithms for plant morphogenesis. In: Sattler R (ed) Theoretical plant morphology. Acta Biotheor 27 (Suppl): Folia Biotheor no 7, pp 37–82

Lindenmayer A (1982) Developmental algorithms: lineage versus interactive control mechanisms. In: Subtelny S, Green PB (eds) Developmental order: its origin and regulation. 40th Symp Soc Dev Biol, 1981. Liss, New York, pp 219–245

Lindenmayer A, Simon N (1979) The problem of theory reduction in genetics. Int Congr Logic, Methodol Philos Sci. Abstracts 9, pp 222–225

Lodkina MM (1983) Features of morphological evolution in plants conditioned by their ontogenesis. Zh Obshsch Biol 44:239–253 (in Russian)

Lorch J (1958) Beitrag zur Kritik der idealistischen Morphologie. Ost Bot Z 105:83–87

Lorenz K (1941) Kants Lehre vom Apriorischen im Lichte der gegenwärtigen Biologie. Blätter Dtsch Philos 15:94–125

Lorenz K (1963) Das sogenannte Böse. Zur Naturgeschichte der Aggression. Borotha-Schoeler, Wien (English translation "On aggression". Bantam Books, New York)

Lorenz K (1971) Knowledge, belief and freedom. In: Weiss P (ed) Hierarchically organized systems in theory and practice. Hafner, New York, pp 231–262

Lorenz K (1973) Die Rückseite des Spiegels. Versuch einer Naturgeschichte menschlichen Erkennens. Piper, München (English translation "Behind the mirror: a search for a natural history of human knowledge". Harcourt, Brace Jovanovich, New York)

Lorenz K (1978) Vergleichende Verhaltensforschung. Grundlagen der Ethologie. Springer, Berlin Heidelberg New York

Lovelock JE (1979) Gaia. Oxford Univ Press, Oxford

Lovtrup S (1981) Introduction to evolutionary epigenetics. In: Scudder GGE, Reveal JL (eds) Evolution today. Univ British Columbia, Vancouver, pp 139–144

Lumsden C, Wilson EO (1981) Genes, mind and cultures. Harvard Univ Press, Cambridge, Mass.

Lumsden C, Wilson EO (1981) Genes, mind and culture. Harvard Univ Press, Cam-Harvard Univ Press, Cambridge, Mass.

Macbeth N (1974) Darwin re-tried. Garnstone, London

Macdonald N (1983) Trees and networks in biological models. Wiley, New York

Mackie JL (1975) The cement of the universe: a study of causation. Clarendon, Oxford

Mainx F (1967) Foundations of biology. Univ of Chicago Press, Chicago

Malacinski GM, Bryant SV (eds) (1984) Pattern formation. Macmillan, New York

268

Marchi E, Hansell RIC (1975) Fuzzy aspects of the parsimony problem in evolution. Math Biosci 23:305−327

Margulis L (1981) Symbiosis in cell evolution. Life and its environment on the early earth. Freeman, San Francisco

Margulis L, Sagan D (1985) L'origine des cellules eucaryotes. La Recherche no 161: 200−208

Martin NH (1967) Rudolf Carnap. In: Encycl Philos 2:25−33

Maruyama M (1974) Hierarchists, individualists and mutualists. Futures, pp 103−113

Maslow AW (1966) The psychology of science. Regnery, Chicago

Masters REL, Houston J (1968) Psychedelic art. A Balance House Book, Grove, New York

Maull N (1977) Unifying science without reduction. Stud Hist Philos Sci 8:143−162

May RM (ed) (1981) Theoretical ecology. Principles and applications, 2nd edn. Blackwell Scientific, Oxford

May RM (1982) Mutualistic interactions among species. Nature 296:803−804

Maynard Smith J (ed) (1982) Evolution now. A century after Darwin. Freeman, San Francisco

Mayr E (1961) Cause and effect in biology. Science 134:1501−1506 (reprinted in Mayr E (1976) Evolution and the diversity of life. Belknap Press of Harvard Univ Press, Cambridge, MA)

Mayr E (1969) Principles of systematic zoology. McGraw Hill, New York

Mayr E (1970) Populations, species and evolution. Belknap Press of Harvard Univ Press, Cambridge, MA

Mayr E (1974) Teleological and teleonomic, a new analysis. In: Cohen RS, Wartofsky MW (eds) Methodological and historical essays in the natural and social sciences. Reidel, Dordrecht (Boston Stud Philos Sci 14)

Mayr E (1976a) Evolution and the diversity of life. Harvard Univ Press, Cambridge, MA

Mayr E (1976b) Is the species a class or an individual? Syst Zool 25:192

Mayr E (1981) Biological classification: toward a synthesis of opposing methodologies. Science 214:510−516

Mayr E (1982) The growth of biological thought: diversity, evolution, and inheritance. Harvard Univ Press, Cambridge, MA

Mayr E, Provine WB (eds) (1980) The evolutionary synthesis. Perspectives on the unification of biology. Harvard Univ Press, Cambridge, MA

McIntosh RT (1980/82) The background and some current problems of theoretical ecology. In: Saarinen E (ed) Conceptual issues in ecology. Pallas Paperback. Reidel, Dordrecht, pp 1−62

McIntosh RT (1984) The background of ecology. Concept and theory. Cambridge Studies in Ecology. Cambridge Univ Press, Cambridge

McMurray WC (1977) Essentials of human metabolism. Harper and Row, New York

Medawar PB (1969) Induction and intuition in scientific thought. Am Philos Soc, Philadelphia

Meeuse ADJ (1981) Evolution of the Magnoliophyta: current and dissident viewpoints. Annu Rev Plant Sci 2:393−442

Melville R (1962/63) A new theory of the Angiosperm flower I. The gynoecium II. The androecium. Kew Bull 16:1−50, 17:1−63

Melville R (1983) Glossopteridae, Angiospermidae and the evidence for angiosperm origin. Bot J Linn Soc 86:279−323

Mendelsohn RS (1979) Confessions of a medical heretic. Warner Books, New York

Mercer EH (1981) The foundation of biological theory. Wiley, New York

Merchant C (1980) The death of nature: women, ecology, and the scientific revolution. Harper and Row, San Francisco

Meyen SV (1973) Plant morphology in its nomothetical aspects. Bot Rev 39:205–260

Meyen SV (1978) Nomothetical plant morphology and the nomothetical theory of evolution: the need for cross-pollination. In: Sattler R (ed) Theoretical plant morphology. Acta Biotheor 27 (Suppl): Folia Biotheor no 7, pp 21–36

Meyen SV (1982) Commentary on Dr. Cusset's paper. Acta Biotheor 31A:87–92

Meyen SV (1984) Basic features of gymnosperm systematics and phylogeny as evidenced by the fossil record. Bot Rev 50:1–111

Miller JG (1978) Living systems. McGraw-Hill, New York

Mishler BD, Donoghue MJ (1982) Species concepts: a case for pluralism. Syst Zool 31:491–503

Mohr H (1977) Lectures on structure and significance of science. Springer, Berlin Heidelberg New York

Mohr H (1982) Principles in plant morphogenesis. Acta Biotheor 31A:93–111

Monod J (1970) Le hasard et la nécessité. Seuil, Paris (English translation: Chance and necessity. Knopf, New York, 1971)

Montagu A (ed) (1980) Sociobiology examined. Oxford Univ Press, Oxford

Morin E (1977) La méthode I. La nature de la nature. II. (1980) La vie de la vie. Seuil, Paris

Nachtigall W (1972) Biologische Forschung. Aspekte, Argumente, Aussagen. Quelle und Meyer, Heidelberg

Nagel E (1961) The structure of science. Harvard, Brace and World, New York

Nagel E (1966) Teleological explanation. In: Canfield JV (ed) Purpose in nature. Prentice-Hall, Englewood Cliffs, NJ, pp 67–88

Nagel E (1979) Teleology revisited and other essays in the philosophy and history of science. Columbia Univ Press, New York

Nakamura H (1964) Ways of thinking of Eastern peoples: India, China, Tibet, Japan. Revised English translation. Wiener PP (ed). East-West Center, Honolulu

Nalimov VV, Meyen SV (1979) Probabilistic vision of the world. 6th Int Congr Logic, Methodol Philos Sci. Abstracts 7, pp 253–257

Nersessian NJ (1982) Why is 'incommensurability' a problem? Acta Biotheor 31:205–218

Newton-Smith WH (1981) The rationality of science. Routledge and Kegan Paul, Boston

Novak M (1971) The experience of nothingness. Harper Colophon Books, New York

Novak VJA (1975) The question of the intraspecific fight for existence and of natural selection from the aspect of biology and philosophy. 5th Int Congr Logic, Methodol Philos Sci. Abstracts 8, pp 21–23

Novak VJA (1982) The principle of sociogenesis. Academia, Prague

O'Brien TP (1982) Growth, divisions and differentiation of cells. In: Smith H, Grierson D (eds) Molecular biology of plant development. Blackwell Scientific, Oxford

Pattee HH (ed) (1973) Hierarchy theory. The challenge of complex systems. Braziller, New York

Patten BC (1975) Ecosystem linearization: an evolutionary design problem. Am Nat 109:529–539

Pauze F, Sattler R (1979) La placentation axillaire chez *Ochna atropurpurea* DC. Can J Bot 57:100–107

Pearce JC (1973) The crack in the cosmic egg. Pocket Books, New York

Pepper SC (1942/70) World hypotheses. Univ California Press, Berkeley

Peters RH (1976) Tautology in evolution and ecology. Am Nat 110:1–12

270

Peters RH (1978) Predictable problems with tautology in evolution and ecology. Am Nat 112:759–762

Peters RH (1983) Ecological implications of body size. Cambridge Univ Press, New York

Pettit P (1979) Rationalization and the art of explaining action. In: Bolten N (ed) Philosophical problems in psychology. Methuen, London

Pittendrigh CS (1958) Adaptation, natural selection, and behavior. In: Roe A, Simpson GG (eds) Behavior and evolution. Yale Univ Press, New Haven, pp 390–416

Plamondon A (1975) The contemporary reconciliation of mechanism and organicism. Dialectica 29:213–221

Platnick NI, Funk VA (1983) Advances in cladistics. Columbia Univ Press, New York

Platt J (1961) Properties of large molecules that go beyond the properties of their chemical subgroups. J Theor Biol 1:342–358

Polanyi M (1964) Science, faith and society. Univ Chicago Press, Chicago

Polanyi M (1968) Life's irreducible structure. Science 160:1308–1312

Pollard JW (ed) (1984) Evolutionary theory: paths into the future. Wiley, New York

Poppendieck H-H (1981) Systematics and evolution of seed plants. Prog Bot 43:188–235

Popper KR (1935) Logik der Forschung. Springer, Vienna (English translation with additions: The logic of scientific discovery. Basic Books, New York, 1959)

Popper KR (1962) Conjectures and refutations. The growth of scientific knowledge. Basic Books, New York

Popper KR (1966) The open society and its enemies, vol 1. Princeton Univ Press, Princeton

Popper KR (1972) Objective knowledge: an evolutionary approach. Clarendon, Oxford

Popper KR (1974) Autobiography. In: Schilpp PA (ed) The philosophy of Karl Popper, vol 1. Open Court, LaLalle, Ill.

Popper KR (1984) Evolutionary epistemology. In: Pollard JW (ed) Evolutionary theory: paths into the future. Wiley, New York, pp 239–256

Popper KR, Eccles JC (1977) The self and its brain. Springer, Berlin Heidelberg New York

Portmann A (1960) Neue Wege der Biologie. Piper, München (translated as "New paths of biology". Harper and Row, New York, 1964)

Portmann A (1965) Beyond Darwinism. Commentary 40:31–41

Portmann A (1974) An den Grenzen des Wissens. Vom Beitrag der Biologie zu einem neuen Weltbild. Econ, Wien (Fischer paperback, 1976)

Presley CF (1967) Quinne WVO In: Encycl Philos 7:53–55

Pribram KH (1982) What the fuss is all about. In: Wilber K (ed) The holographic paradigm and other paradoxes. Shambala, London, pp 27–34

Prigogine I, Stengers I (1979) La nouvelle alliance. Métamorphose de la science. Gallimard, Paris (English version: Order out of chaos: man's new dialogue with nature. Bantam, New York; Shambala, Boston, 1984)

Quine WV (1961) From a logical point of view. 9 essays, 2nd edn. Harvard Univ Press, Cambridge, MA (Also Harper paperback)

Quine WV (1964) On simple theories of a complex world. In: Gregg JR, Harris FTC (eds) Form and strategy in science. Reidel, Dordrecht, Holland, pp 47–50

Radl E (1930) History of biological theories. Oxford Univ Press, Oxford

Radnitzky G (1974) Preconceptions in research: a study. Hum Context 6:1–63

Rajneesh BS (1975a) The mustard seed. A living explanation of the sayings of Jesus from the gospel according to Thomas. Harper and Row, New York

Rajneesh BS (1975b) Tantra. The supreme understanding. Rajneesh Foundation, Poona, India

Rajneesh BS (1975c) My way the way of the white clouds. Grove, New York

Rajneesh BS (1976) Meditation: the art of ecstasy. Harper and Row, New York

Rajneesh BS (1978) Neither this nor that. Reflections on a Zen Master. Sheldon, London

Rensch B (1960) Evolution above the species level. Columbia Univ Press, New York

Rensch B (1970) Biophilosophy. Columbia Univ Press, New York (English translation of "Biophilosophie")

Richmond MH (1979) 'Cells' and 'organisms' as a habitat for DNA. Proc R Soc Lond B Biol Sci 204:235–250

Richmond MH, Smith DC (organizers) (1979) The cell as a habitat. Proc R Soc Lond B Biol Sci 204:113–286

Riddiford A, Penny D (1984) The scientific status of modern evolutionary theory. In: Pollard JW (ed) Evolutionary theory: paths into the future. Wiley, New York, pp 1–38

Rieber RW (ed) (1980) Body and mind: past, present and future. Academic Press, New York

Riedl R (1975) Die Ordnung des Lebendigen. Systembedingungen der Evolution. Parey, Hamburg (English translation: Order in living organisms: a systems analysis of evolution. Wiley Interscience, New York, 1979)

Riedl R (1980) Biologie der Erkenntnis. Die stammesgeschichtlichen Grundlagen der Vernunft. Parey, Berlin (English translation: Biology of knowledge: the evolutionary basis of reason. Wiley, New York, 1984)

Rigler FH (1982) Recognition of the possible: an advantage of empiricism in ecology. Can J Fish Aquat Sci 39:1323–1331

Robinson JO (1972) The psychology of visual illusion. Hutchinson Univ Library, London

Rorty R (1979) Philosophy and the mirror of nature. Princeton Univ Press, Princeton

Rose H (1983) Hand, brain, and heart: a feminist epistemology for the natural sciences. Signs 9:73–90

Rose H, Rose S (1971) The myth of the neutrality of science. In: Fuller W (ed) The social impact of modern biology. Routledge and Kegan Paul, London, pp 283–294

Rose H, Rose S (eds) (1976a) The political economy of science. Ideology of/in the natural sciences. Macmillan, London

Rose H, Rose S (1976b) The radicalisation of science. Ideology of/in the natural sciences. Macmillan, London

Rose S (ed) (1982) Towards a liberatory biology. The dialectics of biology group. Allison and Busby, London

Roszak T (1969) The making of a counter culture. Anchor Books, Doubleday, Garden City, New York

Roszak T (1973) Where the wasteland ends. Anchor Books, Doubleday, Garden City, New York

Rozov MA (1977) Problems of the empirical analysis of scientific knowledge. Nauka, Novosibirsk (in Russian)

Ruse M (1970) Are there laws in biology? Aust J Philos 48:234–246

Ruse M (1971) Narrative explanation and the theory of evolution. Can J Phil 1:59–74

Ruse M (1973) Philosophy of biology. Hutchinson's Univ Library, London

Ruse M (1975) Charles Darwin's theory of evolution: an analysis. J Hist Biol 8:219–241

Ruse M (1977) Is biology different from physics? In: Colodny R (ed) Logic, laws, and life. University of Pittsburgh Press, Pittsburgh, pp 89–127

Ruse M (1979) Sociobiology: sense or nonsense? Reidel, Dordrecht

Ruse M (1982) Teleology redux. Boston Stud Philos Sci 67:299–310

Russell B (1913) On the notion of cause. Proc Aristot Soc (New Ser) 13:1–26

Russell NH (1975) Introduction to plant science. A humanistic and ecological approach. West Publishing, St. Paul

Rutishauser R (1981) Blattstellung und Sproßentwicklung bei Blütenpflanzen unter besonderer Berücksichtigung der Nelkengewächse (Caryophyllaceen s.l.). Diss Botanicae 62. Cramer, Vaduz

Rutishauser R (1983) *Hydrothrix gardneri*: Bau und Entwicklung einer eigenartigen Pontederiacee. Bot Jahrb Syst 104:115–141

Rutishauser R (1984) Blattquirle, Stipeln und Kolleteren bei den Rubieae (Rubiaceae) in Vergleich mit anderen Angiospermen. Beitr Biol Pflanz 59:375–424

Rutishauser R, Sattler R (1985) Complementarity and heuristic value of contrasting models in structural botany I. General considerations. Bot Jahrb Syst 107:415–455

Sahlins M (1977) The use and abuse of biology. An anthropological critique of sociobiology. University of Michigan Press, Ann Arbor, MI

Salmon WC (1980) Causality: production and propagation. In: Asquith PD, Giere RN (eds) Proc 1980 bienniel meet Philos Sci Assoc, vol 2. Philos Sci Assoc, East Lansing, Michigan, pp 49–69

Sattler R (1966) Towards a more adequate approach to comparative morphology. Phytomorphology 16:417–429

Sattler R (1971) Ein neues Sproß-Modell. Ber Dtsch Bot Ges 84:139

Sattler R (1973) Organogenesis of flowers. A photographic text-atlas. Univ Toronto Press, Toronto

Sattler R (1974a) A new conception of the shoot of higher plants. J Theor Biol 47:367–382

Sattler R (1974b) Essentialism in plant morphology. 14th Int Cong History Sci. Proc no 3, pp 464–467

Sattler R (1974c) A new approach to gynoecial morphology. Phytomorphology 24:22–34

Sattler R (1977) On "understanding" organic form. Sophia Perennis 3:29–50

Sattler R (ed) (1978a) Theoretical plant morphology. Acta Biotheor 27 (Suppl): Folia Biotheor no 7

Sattler R (1978b) 'Fusion' and 'continuity' in floral morphology. Notes R Bot Gard Edinb 36:397–405

Sattler R (ed) (1982) Axioms and principles of plant construction. Nijhoff/Junk, The Hague

Sattler R (1984) Homology – a continuing challenge. Syst Bot 9:382–394

Sattler R, Maier U (1977) Development of the epiphyllous appendages of *Begonia hispida* var. *cucullifera*: implications for comparative morphology. Can J Bot 55:411–425

Sattler R, Perlin L (1982) Floral development of *Bougainvillea spectabilis* Willd., *Boerhaavia diffusa* L. and *Mirabilis jalapa* L. (Nyctaginaceae). Bot J Linn Soc 84:161–182

Saunders PT (1980) An introduction to catastrophe theory. Cambridge Univ Press, Cambridge

Schaffer WM, Kot M (1985) Do strange attractors govern ecological systems? Bioscience 35:342–350

Schilpp PA (1974) The philosophy of Karl Popper, 2 vols. Open Court, La Lalle, Ill.

Schrödinger E (1967) What is life? Cambridge Univ Press, Cambridge

Scudder GGE, Reveal JL (eds) (1981) Evolution today. Proc 2nd Int Congr Syst Evol Biol. Hunt Institute for Botanical Documentation, Carnegie-Mellon Univ, Pittsburg

Settle T (1973) Human freedom and 1568 versions of determinism and indeterminism. In: Bunge M (ed) The methodological unity of science. Reidel, Dordrecht, pp 245–266

Sheldrake R (1981) A new science of life. The hypothesis of formative causation. Blond and Briggs, London

Shimony A (1977) Is observation theory-laden? A problem in naturalistic epistemology. In: Colodny RG (ed) Logic, laws, and life. Some philosophical complications. Univ Pittsburgh Press, Pittsburgh, Pa., pp 185–208

Silvertown J (1984) Ecology, interspecific competition and the struggle for existence. In: Birke L, Silvertown J (eds) More than the parts. Biology and politics. Pluto, London, pp 177–195

Simberloff D (1980) A succession of paradigms in ecology: essentialism to materialism and probabilism. Synthese 43:3–39 (also reprinted in: Saarinen E (ed) (1980/82) Conceptual issues in ecology. Reidel, Dordrecht, pp 63–100)

Simon MA (1971) The matter of life. Philosophical problems of biology. Yale Univ Press, New Haven

Simon NPL (1983) The testability of Haeckel's biogenetic law. 7th Int Congr Logic, Methodol Philos Sci. Abstracts, Sect 9

Simpson GG (1949) The meaning of evolution. Yale Univ Press, New Haven

Simpson GG (1961) Principles of animal taxonomy. Columbia, New York

Simpson GG (1964) This view of life. The world of an evolutionist. Harcourt, Brace and World, New York

Simpson GG (1967) The meaning of evolution. A study of the history of life and of its significance for man. Revised edn. Yale Univ Press, New Haven

Sinnott EW (1955) The biology of spirit. Viking, New York

Skolimowski H (1981) Eco-philosophy. Designing new tactics for living. Boyars, Boston

Slobodchikoff CN (ed) (1976) Concepts of species. Benchmark papers in systematic and evolutionary biology, vol 3. Dowden, Hutchinson and Ross, Strodsburg, PA

Sloman A (1978) The computer revolution in philosophy: philosophy, science, and models of mind. Harvester, Hassocks, Sussex

Small E (1979) The species problem in *Cannabis*. Vol 1: Science. Vol 2: Semantics. Corpus (in co-operation with Agriculture Canada and the Canadian Government Publ Centre), Toronto

Sneath PHA, Sokal RR (1973) Numerical taxonomy. The principles and practice of numerical classification. Freeman, San Francisco

Snyder DP (1978) Toward one science. The convergence of traditions. St. Martin's, New York

Sober E (ed) (1984) Conceptual issues in evolutionary biology: an anthology. MIT Press, Cambridge, MA

Sokal RR (1974) The species problem reconsidered. Syst Zool 22:360–374

Stace CA (1980) Plant taxonomy and biosystematics. Arnold, London

Stebbins GL (1974) Flowering plants. Evolution above the species level. Belknap Press of Harvard Univ Press, Cambridge, Mass.

Stebbins GL, Ayala FJ (1981) Is a new evolutionary synthesis necessary? Science 213:967–971

Steeves TA, Sussex IM (1972) Patterns in plant development. Prentice Hall, Englewood Cliffs, NJ (a new edition is in preparation)

Stegmüller W (1969) Probleme und Resultate der Wissenschaftstheorie und Analytischen Philosophie. Band 1. Wissenschaftliche Erklärung und Begründung. Teil 4. Teleologie, Funktionsanalyse und Selbstregulation. Springer, Berlin Heidelberg New York

Stegmüller W (1976) The structure and dynamics of theories. Springer, Berlin Heidelberg New York

Stegmüller W (1977) Collected papers in epistemology, philosophy of science and history of philosophy. I–II. Synthese Library, vol 91. Reidel, Dordrecht Boston

Stent GS (1969) The coming of the golden age: a view of the end of progress. Natural History, Garden City

Stent GS (1978) Paradoxes of progress. Freeman, San Francisco

Stevens J (ed) (1977) One robe, one bowl. The Zen poetry of Ryokan, translated and introduced by J Stevens. Weatherhill, New York

Tait RV (1981) Elements of marine ecology 3rd edn. Butterworth, London

Takhtajan AL (1971) Tectology: history and problems. Systems Res Yearbook USSR Acad Sci, pp 200–277 (in Russian)

Takhtajan AL (1983) Macroevolutionary processes in the history of the plant world. Bot Zh (Leningr) 68:1593–1603 (in Russian, with English summary)

Tart CT (ed) (1969) Altered states of consciousness. Doubleday, Anchor Books, Garden City, New York

Teilhard de Chardin P (1955) Le phénomène humain. Seuil, Paris (translated as "The phenomenon of man". Harper and Row, New York, 1959)

Thom R (1975) Structural stability and morphogenesis. An outline of a general theory of models. Translated from the French edn, as updated by the author, by Fowler DH. Benjamin, Reading, Mass.

Thom R (1983) Mathematical models of morphogenesis. Horwood, Chichester, England

Thompson D'AW (1917) On growth and form. Cambridge Univ Press, Cambridge (2nd edn 1942/79)

Thuillier P (1981) Darwin and Co. Editions Complexe, Brussels

Tivy J (1982) Biogeography. A study of plants in the ecosphere, 2nd edn. Longman, London

Tomlinson PB (1982) Chance and design in the construction of plants. Acta Biotheor 31A:162–183

Tomlinson PB (1983) Tree architecture. The elusive biological property of tree form. Am Sci 71:141–149

Tompkins P, Bird C (1973) The secret life of plants. Harper and Row, New York

Tondl L (1973) Scientific procedures. A contribution concerning the methodological problems of scientific concepts and scientific explanation. Reidel, Dordrecht (Boston Stud Philos Sci 10)

Tonnelat J (1978) Thermodynamique et biologie I. Entropie, désordre et complexité. Maloine, Paris

Tonnelat J (1979) Thermodynamique et biologie II. L'ordre issu du hazard. Maloine, Paris

Toulmin SE (1974) The evolutionary development of natural science. In: Truitt WH, Solomons TWG (eds) Science, technology and freedom. Houghton Mifflin, Boston, pp 106–117

Tremblay M-A (1983) Une perspective holistique dans l'étude de la santé. A holistic approach in the study of health. Trans R Soc Can (4th ser) 21:3–19

Trintscher KS (1973) The non-applicability of the entropy concept in living systems. In: Gray W, Rizzo ND (eds) Unity through diversity, part 1. Gordon and Breach Science, New York, pp 315–340

Troll W (ed) (1926) Goethes Morphologische Schriften. Diederichs, Jena

Troll W (1937) Vergleichende Morphologie der höheren Pflanzen, 1. Teil. Bornträger, Berlin

Troll W (1949) Die Urbildlichkeit der organischen Gestaltung und Goethes Prinzip der „Variablen Proportionen". Experientia (Basel) 5:491–504

Troll W (1954) Praktische Einführung in die Pflanzenmorphologie, 1. Teil. Der vegetative Aufbau. Fischer, Jena

Troll W, Meister A (1951) Wesen und Aufgabe der Biosystematik in ontologischer Beleuchtung. Philos Jahrb (Fulda) 61:105–131

Trungpa C (1973) Cutting through spiritual materialism. Shambhala, Berkeley

Trungpa C (1976) The myth of freedom and the way of meditation. Shambhala, Berkeley

Trungpa C (1981) Journey without a goal: the tantric wisdom of the Buddha. Shambala, Boulder

Tuomi J (1981) Structure and dynamics of Darwinian evolutionary theory. Syst Zool 30:22–31

Tuomi J, Haukioja E (1979) Predictability of the theory of natural selection: an analysis of the structure of the Darwinian theory. Savonia 3:1–8

Tweney RD, Doherty ME, Mynatt CR (eds) (1981) On scientific thinking. Columbia Univ Press, New York

Uexküll J von (1920) Theoretische Biologie. Paetal, Berlin (translated as "Theoretical Biology". Harcourt and Brace, New York, 1926)

Uexküll J von (1957) A stroll through the worlds of animals and men. In: Schiller CH (ed) Instinctive behavior, part 1. Int Univ Press, New York, pp 5–80

Van Laar W, Verhoog H (1971) The relation between philosophy of science and biology exemplified by the problem of explanation. Acta Biotheor 20:274–301

Van Steenis CGGJ (1980) Rheophytes of the world. Sijthoff and Noordhoff, Alphen aan den Rijn

Van Valen L (1976) Ecological species, multispecies, and oaks. Taxon 25:233–239

Varela FJ (1976) On observing natural systems. Francisco Varela in conversation with Donna Johnson. Co-Evol Quart 10:26–31

Varela FJ (1979) Principles of biological autonomy. North Holland, New York

Vester F (1978) Unsere Welt – ein vernetztes System. Klett-Cotta, Stuttgart

Vickery RK Jr (1984) Biosystematics 1983. In: Grant WF (ed) Plant biosystematics. Academic Press, Toronto, pp 1–24

Vollmer G (1975) Evolutionäre Erkenntnistheorie. Hirzel, Stuttgart

Voorzanger B, Van der Steen WJ (1982) New perspectives on the biogenetic law. Syst Zool 31:202–205

Waddington CH (1961) The nature of life. Unwin, London

Waddington CH (ed) (1968–72) Towards a theoretical biology, 4 vols. Edinburgh Univ Press, Edinburgh, Aldine Press, Chicago

Waddington CH (1970) The theory of evolution today. In: Koestler A, Smythies JR (eds) Beyond reductionism. MacMillan, New York

Waddington CH (1975) The evolution of an evolutionist. Cornell Univ Press, Ithaca, New York

Waddington CH (1977) Tools for thought. Cape, London

Wagner WH (1984) A comparison of taxonomic methods in biosystematics. In: Grant WF (ed) Plant biosystematics. Academic Press, Toronto, pp 643–654

Walters RS (1967) Laws of science and lawlike statements. In: Edwards P (ed) Encycl Philos 4:410–414

Waesberghe H Van (1982) Towards an alternative evolution model. Acta Biotheor 31: 3–28

Ward C (1984) Thaipusam in Malaysia: a psycho-anthropological analysis of ritual trance, ceremonial possession and self-mortification practices. Ethos 12:307–334

Wardlaw CW (1965) Organization and evolution in plants. Longmans, London

276

Watts AW (1951) The wisdom of insecurity. A message for an age of anxiety. Vintage Books, New York

Watts AW (1970) Nature, man and woman. Pantheon Books, New York (Vintage Books edn)

Weber R (1978) The enfolding-unfolding universe: a conversation with David Bohm. Revision (1978) pp 24–51 [also in Wilbern K (ed) (1982) The holographic paradigm and other paradoxes. Shambala, London, pp 44–104]

Weberling F (1981) Morphologie der Blüten und der Blütenstände. Ulmer, Stuttgart

Weigel G, Madden AG (1961) Knowledge: its values and limits. Prentice Hall, Englewood Cliffs, NJ

Weil A (1985) A radical look at health and healing. Interview with B Thomson. East West J 15:32–37

Weingartner RH (1967) Historical explanation. In: Encycl Philos 4:7–12

Weiss PA (1967) One plus one does not equal two. In: Quarton GC, Melnechuk T, Schmitt FO (eds) The Neurosciences Rockefeller Univ Press, New York, pp 801–821

Weiss PA (1968) Dynamics of development: experiments and inferences. Academic Press, New York

Weiss PA (1973) The science of life: the living system – a system for living. Futura, New York

White J (1979) The plant as a meta-population. Annu Rev Ecol Syst 10:109–145

White J (1984) Plant metamerism. In: Dirzo R, Sarukhan J (eds) Perspectives on plant population biology. Sinauer Assoc Inc., Sunderland, MA, pp 15–47

White LL (ed) (1968) Aspects of form. A symposium on form in nature and art, 2nd edn. Lund Humphries, London

Whitehead AN (1920) The concept of nature. Cambridge Univ Press, Cambridge

Whitehead AN (1925) Science and the modern world. Macmillan, New York

Whitehead AN (1929) Process and reality. Cambridge Univ Press, Cambridge

Whitmore TC (1976) Natural variation and its taxonomic treatment within tropical tree species as seen in the Far East. Linn Soc Symp Ser 2:25–34

Whittaker RH (ed) (1973) Ordination and classification of communities. Junk, The Hague

Whittaker RH, Gauch HG Jr (1978) Evaluation of ordination techniques. In: Whittaker RH (ed) Ordination of plant communities. Junk, The Hague

Wickler W (1972) The biology of the ten commandments. McGraw Hill, New York

Wiener N (1961) Cybernetics, 2nd edn. MIT Press, Cambridge

Wilber K (ed) (1982) The holographic paradigm and other paradoxes. Exploring the leading edge of science. Shambala, Boulder

Wilber K (ed) (1984) Quantum questions. Mystical writings of the world's great physicists. New Science Library, Shambala Publ, Boulder, Co.

Wiley EO, Brooks DR (1982) Victims of history – a nonequilibrium approach to evolution. Syst Zool 31:1–24

Wilhelm H (1960) Change. Eight lectures on the I Ching. Translated by Baynes CF. Princeton Univ Press, Princeton (Harper Torchbook paperback 1973)

Williams GC (1966) Adaptation and natural selection. Princeton Univ Press, Princeton

Williams MB (1970) Deducing the consequences of evolution: a mathematical model. J Theor Biol 29:343–385

Williams MB (1973) The logical structure of natural selection and other evolutionary controversies. In: Bunge M (ed) The methodological unity of science. Reidel, Dordrecht, pp 84–102

Williams MB (1979) Circularity at the core of a theory: deep tautology versus mere tautology. 6th Int Congr Logic Methdol Philos Sci. Abstracts 9, pp 226–229

Williams MB (1982) The importance of prediction testing in evolutionary biology. Erkenntnis 17:291–306

Wilson EO (1975) Sociobiology: the new synthesis. Belknap Press of Harvard Univ Press, Cambridge, Mass.

Wilson EO (1978) On human nature. Harvard Univ Press, Cambridge, Mass.

Wimsatt WC (1976) Reductive explanation: a functional account. Boston Stud Philos Sci 32:671–710

Wimsatt WC (1980) Randomness and perceived randomness in evolutionary biology. Synthese 43:287–329

Wittgenstein LJJ (1922) Tractatus logico-philosophicus. Routledge und Kegan Paul, London

Wolvekamp HP (1982) The animal as a pluricausal system. Acta Biotheor 31:29–43

Woodfield A (1976) Teleology. Cambridge Univ Press, Cambridge

Woodger JH (1937) The axiomatic method in biology. Cambridge Univ Press, Cambridge

Woodger JH (1952) Biology and language: an introduction to the methodology of the biological sciences including medicine. The Tarner Lectures for 1949–50 Cambridge Univ Press, Cambridge

Woodger JH (1959) Studies in the foundations of genetics. In: Henkin L, Suppes P, Tarski A (eds) The axiomatic method in biology. North-Holland, Amsterdam, pp 408–428

Woodger JH (1967) Biological principles. Reissued (with a new introduction). Humanities, New York

Woozley AD (1967) Universals. In: Edwards P (ed) Encycl Philos 8:194–206

Wright L (1976) Teleological explanation. Univ California Press, Berkeley, CA

Wuketits FM (1978) Wissenschaftstheoretische Probleme der modernen Biologie. Duncker and Humblot, Berlin

Wuketits FM (1981) Biologie und Kausalität. Biologische Ansätze zur Kausalität, Determination und Freiheit. Parey, Berlin

Wuketits FM (1983) Biologische Erkenntnis: Grundlagen und Probleme. Fischer, Stuttgart

Wuketits FM (ed) (1984) Concepts and approaches in evolutionary epistemology. Reidel, Dordrecht

Yarranton GA (1967) Organismal and individualistic concepts and the choice of methods of vegetation analysis. Vegetatio 25:113–116

Young JF (1969) Cybernetics. American Elsevier, New York

Zadeh L (1965) Fuzzy sets. Inf Control 8:338–353

Zadeh L (1971) Similarity relations and fuzzy orderings. Inf Sci 3:177–200

Zeeman EC (1976) Catastrophe theory. Sci Am 234:65–83

Zimmermann W (1968) Evolution und Naturphilosophie. Duncker and Humblot, Berlin

Zukav G (1979) The dancing Wu Li Masters. Bantam Books, New York

*For further references see:*

Blackwell RJ (compiler) (1983) A bibliography of the philosophy of science 1945: 1981. Greenwood, Westport Conn., London

Grene M, Burian R (1983) Philosophy of biology in the philosophy curriculum. A publication developed at the 1982 Summer Institute of Philosophy of Biology. The Council for Philosophical Studies, San Francisco State University, San Francisco. See also the new journal *Biology and Philosophy*

# Subject Index

Synthetic theory of evolution   192–
    198
System   134, 139, 168, 218–219
Systems model of scientific methodo-
    logy   31–35, 106–109
Systems theory of evolution   196
Systems thinking   3, 139, 163

Tao   48, 194
Tautology   197–198
Teleology   151–180
    developmental   153
    external   154
    fuzzy   161
    human   158–160
    instrumental   155
    internal   154
Teleomatic processes   155
Teleonomy   153–154, 162, 227
Tenacity, principle of   32–34
Tender-minded   187
Term   78–80, 120
    primitive   81, 120, 198
Theory   10, 101–103, 111, 221
    semantic view of   40–41, 111, 113
Theory change   32
Theory coexistence   109
Theory-ladenness   63
Theory-reduction   221–224
Theory replacement   222–223
Thinking
    capitalist   203
    causal   125
    dualistic   205
    either-or   146
    formist   247
    hierarchical   96, 105, 135
    linear   128
    morphological   124
    network   131, 135–139
    systems   3, 139, 163, 211
Thought   70
Thought-object   119
Tolerance   22, 110
Tough-minded   187
Trance   137
Transformation   170
Trigger   132
Trinity   106
Truth   75, 117–118, 249–252
    causal-adjustment theory of   250

coherence theory of   252
correspondence theory of   250
operational   251
qualitative confirmation theory of
    251
successful working theory of   251
verified hypothesis theory of   251
Type   85, 121

Umwelt   14, 62
Uncertainty   21, 22, 23, 141
    existential dimension of   23
    social dimension of   22
Uncertainty principle   140–141
Understanding   57–58, 116
Unification of theories   39
Uniformitarianism   184
Uniqueness   154, 184, 186–188, 227
Units   219–220
Unity   15, 68, 80, 89, 121, 133, 146,
    220
Universality   112
Unnamable, the   48, 51, 80, 86, 113,
    235

Values   202–206
    human   202–206
Values of science   32, 106, 202–206
Variability   57, 113
Variable   133
Variation
    clinal   98
Verification   22
Vital force   153, 212
Vital principle   212–214
Vitalism   212–215

Ways of thinking   241
Whole   2, 69, 118, 135, 175, 194, 205,
    227, 229, 245, 246
Whole plant physiology   136
Wholeness   133, 220, 236
Wisdom   50, 255
    crazy   50
World hypotheses   241–252
World views   241–252

Yoga   147

Zen   70, 118, 186, 220